P9-AGG-274

A History *of* Molecular Biology

A History *of* Molecular Biology

MICHEL MORANGE

Translated by Matthew Cobb

HARVARD UNIVERSITY PRESS
Cambridge, Massachusetts, and London, England / 1998

Printed in the United States of America

Originally published as *Histoire de la biologie moléculaire,* © 1994, Editions La Découverte.

Publication of this book has been aided by a grant from the French Ministry of Culture.

Library of Congress Cataloging-in-Publication Data

Morange, Michel.
[Histoire de la biologie moléculaire. English]
A history of molecular biology / Michel Morange ; translated by
Matthew Cobb.
p. cm.
Includes bibliographical references (p.) and index.
ISBN 0-674-39855-6 (alk. paper)
1. Molecular biology—History. I. Title.
QH506.M7213 1998
572.8′09—dc21 97-47158

Contents

A History *of* Molecular Biology

Introduction

Barely a day goes by without the media's focusing on another new development in biology—gene therapy, the human genome project, the creation of new varieties of animals and plants by genetic engineering, and even the possibility of cloning a human being. Naturally enough, the public is fascinated. These possibilities have all come about because of molecular biology, which developed in the middle of the twentieth century.

Molecular biology is not merely the description of biology in terms of molecules—if this were the case, it would include not only biochemistry, but also all those nineteenth-century studies in chemistry and physiology that led to the characterization of biological molecules. With such a broad definition, even Pasteur would have been a molecular biologist![1] Rather, molecular biology consists of all those techniques and discoveries that make it possible to carry out molecular analyses of the most fundamental biological processes—those involved in the stability, survival, and reproduction of organisms.

Molecular biology is a result of the encounter between genetics and biochemistry, two branches of biology that developed at the beginning of the twentieth century. These two disciplines had each clearly defined the object of their research: the gene for genetics, proteins and enzymes for biochemistry. Molecular biology emerged when the relation between these two objects became clearer. Scientists identified the gene as a macromolecule (DNA), determined its structure, and described its role in protein synthesis.

Strictly speaking, molecular biology is not a new discipline, but rather a new way of looking at organisms as reservoirs and transmitters of information.[2] This new vision opened up possibilities of action and intervention that were revealed during the growth of genetic engineering.

If the content and the history of molecular biology are difficult to define, it is, by contrast, relatively easy to describe the period in which this molecular revolution took place. The new conceptual tools for analyzing biological phenomena were forged between 1940 and 1965. The consequent operational control was acquired between 1972 and 1980. This book covers the molecular revolution in its entirety. Molecular biology and genetic engineering are too intimately linked for their histories to be separated; genetic engineering cannot be understood without molecular biology, but it was genetic engineering that highlighted the importance of the conceptual changes that molecular biology had produced.

Molecular biology was born when geneticists, no longer satisfied with a quasi-abstract view of the role of genes, focused on the problem of the nature of genes and their mechanism of action. It was also a result of biochemists' trying to understand how proteins and enzymes—essential agents of organic specificity—are synthesized and how genes intervene in this process.

The end of this history is more difficult to discern. We still live within the molecular paradigm,[3] and contemporary biologists work in the conceptual frameworks established more than thirty years ago. This book closes with the explosion of knowledge that followed the development of the principal methods of genetic engineering. Only some of the discoveries made possible by genetic engineering will be described, simply to illustrate the nature of current research in molecular biology.

An exception to this general approach will be made for the 1983 discovery of a technique for amplifying DNA—"PCR"—which is emblematic of the molecular revolution. This discovery had its origins in the theoretical framework developed in the 1950s, and in the experimental tools devised in the 1970s. Better than any other example, it shows the effectiveness of the theoretical and practical tools forged by biologists in the second half of the twentieth century.

A major problem in writing the history of the molecular revolution in the life sciences is the sheer mass of documentation available. Many of the participants in this revolution have written their own accounts. There are also a large number of studies of these discoveries by scientists, historians, and philosophers.[4] Two books in particular, with very different approaches, have made a major contribution to the history of molecular biology. The English historian Robert Olby has written a detailed account of the "path" that led to the discovery of the double helix structure of DNA.[5] Horace F. Judson interviewed more than one hundred of the most important participants in the history of molecular biology and reconstituted the technical and conceptual debates that surrounded its birth.[6]

The aim of this work is not to repeat what has already been done well by Olby and Judson. On the one hand, this history of molecular biology describes the development of genetic engineering and its first results. Up to now, this story has been told only in a fragmented manner.[7] The history of genetic engineering makes it easier to appreciate the originality of the new view of life that constitutes molecular biology. On the other hand, Olby's and Judson's books have been complemented or criticized in a series of articles on the history of science that have been published in specialist journals. These studies have not yet been gathered together in a general history of molecular biology. Rather than simply describe the discoveries, these articles try to explain how and why they took place. Many of these studies are included in this book, an important part of which deals with the question of the "origin" of molecular biology, which the historian Lily Kay has also recently addressed.[8]

A contemporary science poses a number of specific problems for the historian. There are many sources: in addition to the traditional material (books, articles, laboratory notebooks), there are the scientists' own accounts of their discoveries, which can take the form of autobiographical books or articles, interviews, and oral accounts. This abundance of sources does not always make for clear history. Dominique Pestre has shown that extreme care is necessary in dealing with oral accounts, which more often than not are reconstructions rather than historical documents.[9] In the case of molecular biology, the authors of these autobiographical accounts have understandably tended to justify their own role, or the role of their discipline, in the development of molecular biology. Most of these accounts have been written while their authors still occupied important positions and played major roles in the definition of scientific policy. Consciously or unconsciously, such accounts are often marked more by strategic motivations than by a concern for historical truth.

Another difficulty in arriving at a historical analysis of recent scientific developments is that historians tend to interpret the past through the eyes of the present. This danger, intrinsic to any historical analysis, is all the greater when the words and techniques are the same as those used today. The historian runs the risk of projecting a contemporary mind-set onto past experiments. This risk is even greater if the history is written by a scientist who believes that a particular discipline's past is merely an anticipation of its present.

My aim was to write a book on the history of science that could be read by the general public. Too many articles and books require the reader to understand the object of the history before reading the first word. In this book, a substantial amount of space is devoted to explaining discoveries, and in particular (to the extent that this is possible) to describing the techniques involved. Because general readers may find some of the terminology employed in Parts 2 and 3 difficult, I have added an appendix that summarizes the key results of molecular biology.

I also wanted to write a history that was as complete as possible. Many previous accounts have emphasized one or another research school. I have tried to provide a balanced presentation of the disciplines that contributed to the development of molecular biology,[10] in particular by emphasizing the role of biochemistry.

Finally, this book contains biographies of several scientists who played major roles in the birth of molecular biology. This is the reflection of the rich biographical and autobiographical material available. A scientific biography has an added interest: it enables the historian to step outside the framework of a purely internal history and to outline the role of external factors, such as the cultural context in which molecular biology was born. The occasionally chaotic path followed by the founders of the new science, their movements between disciplines or between countries, all contributed to the cultural mixture that gave rise to the new biology.[11]

That's a lot to ask of one book! For this reason, I have not been able to devote as much space as I would have liked to each of the developments in molecular biology. Each chapter is devoted to a different theme: a discovery, a research group, or a particular historical question. If this thematic presentation has the advantage for the reader of making each chapter independent, it has the disadvantage of not following the chronological order of events.

It is normal practice in the history of science for an author to set out his or her strategy. Although I will not try to situate myself in the complex world of different schools and currents, it is worth making a brief statement of aims, given that the history presented here raises the fundamental question of whether there has been a "molecular revolution" or a slow evolution of biology toward the study of biological phenomena at the molecular level. These two apparently contradictory interpretations—revolution or evolution—are in fact both possible and complementary.

Fernand Braudel has shown that history follows different rhythms and tempos, and can be separated into several currents.[12] To my knowledge, this model has never been applied to the history of science, but the history of molecular biology becomes clearer when it is seen as a result of three parallel but different histories.

The longest time frame is that of reduction—the reduction of biology to physics and mechanics, which began in the seventeenth century,[13] and the reduction of biology to chemistry, which began in the nineteenth century. The history of this reductionist current is complex, with steps forward and backward, but it formed a powerful wave that, in the middle of the twentieth century, swept biology to the feet of the structural chemists.

On this long time frame is superposed the shorter history of the various biological disciplines. The key events here are the birth of biochemistry and genetics at the beginning of the twentieth century. Seen from this point of view, molecular biology is the fruit of the convergence of these two disciplines, and also the beginning of a reconciliation between heredity and development, between genetics and embryology.

Finally, these slow transformations form the backdrop to the history of events, the history that decided experiments and theories. The events that influenced the birth of molecular biology do not belong only to the history of science; indeed, the migration of many scientists to Great Britain and the United States prior to the Second World War, the growing communication needs, and the birth of computing, which was linked to them, all gave molecular biology its current form.

Viewing molecular biology as the result of three different histories frees us from the sterile counterposition of evolution and revolution: what seems to be a revolution at one level may be revealed as an evolution when history is seen on a longer scale. It becomes clear that the influences that shaped molecular biology converged in the unfolding of these histories.

Whereas the reductionist current is linked with the birth of modern science in the sixteenth century, the chemical form it later adopted was the product of the nineteenth-century development of chemistry in both the fundamental and the applied spheres.

The intermediate history of disciplines is the most appropriate for analyzing the confrontation between conceptual history and social history. The history of genetics is the history of the discovery of genes, but it is also the history of geneticists and their strategies. The same is true for biochemistry, and for the other disciplines that participated in the birth of molecular biology, such as virology and bacteriology.

Finally, the event-based history can be seen throughout this confrontation between the internal developments of biology (for example, the fact that Avery, almost despite himself, discovered that genes are made of DNA) and the external influences that led the physicists to be the tutors of the new discipline. Riding on the crest of these three waves throughout the second half of the twentieth century, molecular biology has transformed biology.

The final "methodological" point that should be made is to repeat the conclusion of Michel Callon and Bruno Latour's brief but lucid summary of the problem:[14] what the history of science lacks most is, strangely enough, the history of science and its controversies. That is, we require a painstaking, "anthropological" description of the behavior of researchers when they elaborate theories and models, but also, and above all, when they do experiments.

Even in a discipline like molecular biology, whose history has been so well studied, large areas remain barely explored, or are even untouched. For example, whereas the history of the discovery of the double helix has been described over and over again, the discovery of DNA polymerase, the enzyme that duplicates DNA, is barely mentioned even in the most thorough historical studies. More than any other discipline, the history of science cannot avoid "fashion"[15]—all the more so because historians of science are often external to the scientific medium that they study. Relatively unfamiliar with what is—or was—done on a day-to-day basis in biology laboratories, such writers tend to select the most spectacular and highly publicized aspects of science. Refusing to take the usual path in studying the history of science—reading scientific publications and understanding the possibilities and limits of the techniques employed—they prefer to interview the "stars" and to pore over their correspondence. This attitude, frequently justified by long methodological explanations,[16] often goes against the proclaimed objective: it reinforces an intellectual and elitist vision of science.

Too great a distance from the material also explains the biased interpretation that some historians have made of David Bloor's principle of symmetry,[17] which argues that historians should analyze success and failure in the same way; according to Bloor, the same analytical framework should be employed to account for a theory that succeeded and for a theory that failed.

This is an excellent principle, because the natural tendency of scientists is to consider that the best theory was necessarily the one that triumphed, the one that history has retained. But it has sometimes been used to relativize science, to suggest that all theories are equal, and that those that "lose" have been eliminated for "extrascientific" reasons.[18]

This position does not take into account how science works and the fundamental role played by experimental constraints.[19] Scientists can elaborate all the models and theories they like, but they cannot oblige their experiments to "work."[20] Even a short stay in a biology laboratory will show that biological material is difficult to work with, and that experiments that fail have an important place in everyday laboratory life. Scientific knowledge is "constructed" and scientists are free to define their strategy and to elaborate their models, but only within the narrow limits left to them by the experimental systems they use.

Above all, I have tried to be precise in historical terms: to describe, as faithfully as possible, the known and unknown aspects of the history of molecular biology.[21] Whatever the value of the interpretations put forward here, this book and the historical information it contains will enable others to take us further in the understanding of the molecular revolution in biology.

PART I

The Birth of Molecular Biology

CHAPTER 1

The Roots of the New Science

At the beginning of the twentieth century, biochemistry replaced physiological chemistry.[1] Biochemistry provided medicine with scientific methods of diagnosis; as a fundamental science, it attempted to understand the ways in which molecules are transformed within organisms.

The first biochemical experiment took place in 1897, when Edward Büchner, a German chemist, succeeded in reproducing sugar fermentation *in vitro,* using a cell-free yeast extract. This was particularly significant because forty years earlier Pasteur had argued that fermentation represented the "sign" or "signature" of life.[2]

Biochemistry developed in two directions: on the one hand it studied the transformation of molecules (in particular, sugars) within organisms; on the other, it characterized proteins and enzymes, essential constituents of life and the agents of the molecular transformations that interested biochemists.

The first half of the twentieth century was an important period for biochemistry, marked by the decoding of the major metabolic pathways and cycles—the glycolysis pathway, the urea cycle, Krebs's cycle, and so on—and by a large number of studies on cell respiration. At the same time, advances in physical chemistry led to the development of media that would permit the *in vitro* study of enzyme activity.[3] A quantitative scale of acidity ("pH") was developed, as were "buffer solutions" that reproduced the properties of the intracellular medium. These developments laid the basis for the study of the fundamental principles of the kinetic activity of enzymes—enzymology. The purification and crystallization of enzymes made it possible to study their structure.

In biochemistry, the first two decades of the twentieth century were dominated by "colloid" theory.[4] At the interface of chemistry and biology, this theory proclaimed the existence of a new state of matter—the "colloidal" state—with physical and chemical properties that were characteristic of organisms but could nevertheless be studied by physics and chemistry. The boundaries of this theory were ill-defined, and a number of studies carried out within its framework are now considered classics in physical chemistry. Despite the fact that colloid theory has since sunk into oblivion, it was extremely important; indeed, several Nobel Prizes were awarded for work on colloids.

We now know, however, that this theory was completely wrong on a number of points. One of its key postulates was that colloids were formed when low molecular weight elementary molecules were grouped together. Supporters of colloid theory thought that large molecules could not exist, and that there could only be aggregates of smaller molecules.

A battle took place between supporters and opponents of colloid theory over the measurement of molecular mass. The first stage of this battle was the development of appropriate measurement techniques. The isolation of crystallized proteins and enzymes and the development of X-ray diffraction images of these crystals showed that the components of organisms have well-defined structures, something that was incompatible with colloid theory.[5]

Colloid theory was gradually replaced by macromolecular theory. The term "macromolecule" was introduced in 1922 by the German chemist Hermann Staudinger to describe high-mass molecules in which atoms are held together by strong bonds.

Another concept that is important for understanding the history of biochemistry and the particular contribution of molecular biology is "specificity." "Specificity" refers to the ability of enzymes to recognize the chemical structure of the molecules on which they act (their substrate). This concept, which was thought to be characteristic of biological molecules, was omnipresent in the biology of the first half of the twentieth century.[6] The first clear reference to it was made in 1890 by the German chemist Emil Fischer, who had undertaken an extensive study of proteins. To illustrate specificity, Fischer used the metaphor of a lock and key—the enzyme interacts with its substrate like a key with its lock.

The idea that all biological molecules had a chemical specificity developed from research on enzymes and found its most striking development in immunology. Immunologists quickly made an analogy between the enzyme-substrate interaction and the interaction between antigens and antibodies.

The chemical study of antibodies—immunochemistry—developed in the first half of the twentieth century under the impetus of Karl Landsteiner.[7] Landsteiner, an Austrian immunologist, worked at the Rockefeller Institute in New York. He injected animals with various molecules and studied how antibodies were formed against them. The results were impressive: whatever the chemical nature of the molecules, the organism was capable of producing antibodies against them, provided that they were coupled with a carrier macromolecule.

Landsteiner made small molecular modifications to the injected substances and showed that the animal could recognize these variations and synthesize antibodies that react specifically with the new molecules. These results showed that specificity of recognition is an intrinsic property of life. Immunochemistry, though a discipline in its own right, proved to be a tool for studying the composition of an organism. If an animal is injected with proteins extracted from other organisms, antibodies directed specifically against a given protein from a given organism are produced. The specificity of immunological recognition thus reveals the specificity of the constituents of an organism.

Molecules found in organisms have their own particular forms and properties that enable them to interact specifically with other organic constituents—antibodies or substrates. From an experimental point of view, a biological constituent is specific if an animal can produce antibodies directed "specifically" against it.

Between 1936 and 1940, the concept of specificity changed substantially, evolving from a biological concept into a stereochemical one. This rapid evolution, essential for organic properties to be reduced to the physicochemical level, was the result of Landsteiner's meeting with the American chemist Linus Pauling.[8] Landsteiner wanted to find a chemical explanation for the specificity shown by the antibodies that were produced against the molecules he injected. For Pauling—who wanted to study biological molecules—Landsteiner's results were excellent material for characterizing the chemical bonds responsible for the interaction between antigens and antibodies.

Pauling was already well known for having adapted the concepts of quantum mechanics to the study of molecules.[9] The theoretical work of the Austrian physicist Erwin Schrödinger, together with Walter Heitler and Fritz London's application of quantum mechanics to the hydrogen molecule, had shown that the formation of chemical bonds could be explained by quantum mechanics and could be predicted from the structure of the atoms involved in the bond. The calculations were so difficult, however, that the new quantum theory could not be applied to complex molecules. Pauling simplified these calculations and showed, on the basis of many examples, that quantum mechanics could explain the existence and the characteristics of the chemical bond. He subsequently extended this work to a number of other molecules where the calculations gave results for bond length and strength that did not match the observed values. Pauling suggested that this difference resulted from the fact that the molecules "resonated" between several different structures. The observed difference was the direct translation of this equilibrium between two forms, the result of the resonant energy that was thus created. Pauling used this insight to reinterpret previous experimental data, in particular those from crystallographic studies. His semi-empirical approach, which continually shifted between structural studies and simple theoretical rules derived from the principles of quantum mechanics, gave chemistry a new form.

Pauling's personality, his charisma, and his gift for teaching also played an important part in this transformation.

This approach had previously allowed Pauling to distinguish strong covalent bonds from "weak" bonds. Strong bonds are classically called "chemical bonds"—they arise when two atoms share electrons. Weak bonds—hydrogen or ionic bonds—are produced by the partial sharing of electrons; despite their characterization as "weak," they play an important role in biology.

Landsteiner's data were a striking demonstration of the importance of weak bonds in molecular interactions. Pauling explained the specificity of the antibody-antigen interaction by the formation of a certain number of weak bonds, in particular, hydrogen bonds between the antigen and the antibody. Weak bonds are formed only between atoms situated close to each other. The existence of a large number of weak bonds showed that antigens and antibodies have complementary structures and fit into one another, thus confirming the model proposed half a century earlier by Emil Fischer. With Pauling, the concept of specificity acquired real chemical credentials and became understood in terms of stereospecificity, structural complementarity, and collections of weak bonds.[10]

Applying the concept of weak bonds to protein structure, Pauling studied the denaturation of proteins by heat, acids, bases, or chemicals such as urea. In 1936 he arrived at a correct interpretation of denaturation:[11] it involves not the breaking of a covalent bond in the molecule nor the separation of a colloidal aggregate, but the loss of a set of hydrogen bonds necessary for the stabilization of the protein's three-dimensional structure.

Pauling thus played an essential role in reducing specificity to physics and chemistry. The notions of weak bonds and structural complementarity are still the basis of our understanding of macromolecular interactions. These principles govern the structure and functioning of all forms of life.

The history of genetics is no less compelling than that of biochemistry. In 1866 the Moravian monk Gregor Mendel first formulated the "laws of heredity" that were rediscovered in 1900 by Hugo de Vries, Erich von

Tschermak, and Carl Correns.[12] But it was several years before the laws assumed the form in which they are known today. Only in 1909 did Wilhelm Johannsen introduce the fundamental distinction between genotype and phenotype. "Genotype" refers to the collection of pairs of factors transmitted over generations, one of which is contributed by the father, the other by the mother. These factors are called "genes." "Phenotype" refers to the totality of adult characters for which there are several forms that can be easily distinguished.

The expansion of genetics was closely linked to the choice of the fruit fly *(Drosophila)* as a model organism. The American biologist Thomas H. Morgan and his collaborators at Columbia University chose *Drosophila* because of its rapid reproduction rate.[13] This makes it particularly easy to study the genetics of its characters. In 1910 Morgan discovered that some genes were transmitted differently depending on the sex of the fly that carried them, in the same manner as the "sex chromosomes." At a stroke, this confirmed both the role of the sex chromosomes in sex-determination and the hunches of Theodore Boveri and Walter S. Sutton that genes were carried on chromosomes.

The recombination of genes carried by a pair of chromosomes was explained by the exchange of chromosome fragments that takes place during the formation of sex cells. The closer together the genes, the rarer the recombination. Recombination allowed Morgan's group to order genes on the chromosome and thus to establish chromosome maps.

The choice of *Drosophila* proved to be particularly astute. The salivary glands of the fruit fly contain giant chromosomes that can easily be seen under the microscope, where they appear as a series of alternating light and dark bands. Changes observed in these bands when genes were altered were used to draw up a physical map of the genome, which could be superimposed on the genetic map established on the basis of recombination frequencies. This major discovery confirmed that genes were physical entities and that they were localized on chromosomes.

Geneticists not only characterized genes and localized them on chromosomes, but also described mutations and chromosome alterations and analyzed their effects on organisms' morphological and physiological characters. They showed that some types of physical or chemical treatments (such as X-rays) could increase the mutation rate.

The genetic analysis of *Drosophila* became a model for the genetic analysis of other organisms—both plants and animals. All this work confirmed the validity of the genetic analysis of reproduction and helped to establish genetics as a science.

The essential role of genetics in the birth of molecular biology can be explained as follows:

- Genetics became a leading biological science because of its elegant models and its influence on agronomy. In both the United States and Great Britain, farmers and seed merchants contributed to the development of research and the creation of genetics institutes, aided by public funding.[14] Genetics was attractive because of its practical successes—its models provided an effective guide to both applied and fundamental research. Genetics developed much faster in the United States than in France or Germany because the flexibility of the American university system, unlike the French or German systems, enabled it to rise to the challenge of the new science.[15]
- Genetics rapidly developed into a separate discipline that had little contact with other branches of biology. Far from being a handicap, this institutionalized isolation benefited the new science. Genetics made its greatest steps forward where it developed its own particular character as a new subject—in particular in the United States. In those countries where genetics was not recognized as an independent discipline (for example, in Germany[16]), genetic research was done in departments of physiology or biology, and did not develop to the same extent.

Thus the isolation of genetics was necessary for its development as a discipline. To understand this, we have to go back to Mendel's discovery of the "laws of heredity." Mendel's results had to wait forty years before they were understood and accepted. This period was necessary for heredity to become separated from embryology, a process that was essential for the birth of genetics.

The study of reproduction is the study of how two organisms give rise to a third, similar, organism. It involves examining two intimately linked questions: the development of the organism from egg to adult, and the similarities between the organism and its parents. The success of genetics shows that the separation of these questions—the separation of heredity and embryology—was necessary at least for a time.[17]

Because of its intellectual isolation, many biologists considered genetics a formal science removed from reality. Geneticists studied genes without apparently being interested either in the way they worked or in their chemical nature. Hermann J. Muller was the only one of Morgan's students to show an immediate interest in these questions. For Muller, genes were the "basis of life," the place where the "secrets" of life could be found. In 1927 Muller discovered the mutagenic effect of X-rays. Just as Ernest Rutherford's 1919 transmutation of chemical elements had opened the road to an understanding of the atomic nucleus, so Muller was convinced that his 1927 experiment would help unravel the mysteries of genes.[18] Muller was particularly interested in the work of Max Delbrück, who was trying to define the properties of genes by studying variations in mutation rate as a function of radiation energy (see Chapter 3).

Nevertheless, there are two reasons—one experimental, the other theoretical—that geneticists took only a limited interest in such research. The first reason was that in the 1930s direct chemical study of chromosomes had shown that they were composed of nucleic acids (deoxyribonucleic acid, or DNA) and proteins. This result was rapidly accepted by biologists,[19] but did not imply that the two components were equally important. Genes were thought to be composed of proteins, with nucleic acids serving as a material support or as energy reserves. The chemical nature of genes was not considered important for the simple reason that most geneticists thought the problem at least partly solved.

The second reason geneticists were not immediately interested in studying how genes worked is that they already had an idea of the role of genes in the cell. In a 1945 conference Muller described genes as "guides . . . relatively invariable . . . that serve as a frame of reference." Taken together, the genes in an organism were thought to form "a relatively stable controlling structure to which the rest is attached."[20] Geneticists may have considered genes the basis of all organisms, but this did not mean that they distinguished genes from other components of the organism; the study of the chemistry of genes was linked to the study of the other components.[21]

Geneticists did, however, believe that genes had a fundamental and unique property: self-replication. This capacity was consubstantially linked to genes themselves and could be studied independent of their functioning. Geneticists therefore placed much more emphasis on the self-replicating function of the gene than on its action in the cell. Self-replication was particularly intriguing and gave rise to numerous models. Some scientists related it to the pairing of identical chromosomes, arguing that like attracted like in both cases.[22]

Pascual Jordan proposed a model of gene replication based on the principles of quantum mechanics (of which he was one of the founders),[23] according to which genes replicated by attracting their constituent elements from the surrounding medium, the interaction between two identical elements being favored by the existence of a long-range resonant energy. In 1940, Pauling and Delbrück refuted Jordan's model by showing that it was incompatible with the results of quantum mechanics. They suggested that the force of attraction between identical molecules stemmed from the existence of a structural complementarity at the submolecular level.[24]

Pauling and Delbrück's paper may seem to anticipate the complementary double helix structure of DNA and its replication. In fact, such an interpretation would be anachronistic and would give the paper more weight than even its authors attributed to it. Pauling himself did not refer to the paper when he tried to determine the structure of DNA in the 1950s. This paradox shows that the theoretical study of the self-replicating properties of genes was not an operative research program.

Genetics was thus separated from other fields of biology such as embryology or biochemistry, whereas its relations with evolutionary biology were complex. Genetic models were generally received with little enthusiasm; many evolutionary biologists (both Darwinians and neo-Lamarckians) had learned to reason in terms of progressive evolution and continuous variations that were accumulated over generations. The mutations discovered by the geneticists were far larger than the microvariations proposed by many evolutionary theorists. Furthermore, there were close links between the Darwinians and the biometricians, who measured the variability of hereditary characters and their transmission from generation to generation. Although they studied the same objects as the geneticists, the biometricians had a completely different theoretical approach. The traits they studied (such as size or weight) are today known to be polygenic characters (produced by the action of a number of genes), which makes it particularly difficult to study how they are transmitted.

The reconciliation of genetics and Darwin's theory of evolution was a slow process, in which the mathematical geneticists R. A. Fisher, J. B. S. Haldane, and Sewall Wright played an important part. Population genetics emerged in the 1930s as a result of their debates and writings, and was followed by the "evolutionary synthesis," which took place under the influence of Theodosius Dobzhansky.[25] The neo-Darwinian synthesis explains evolution by a combination of two phenomena:

- the existence of spontaneous small-scale genetic variations (mutations); and
- the elimination of less fit individuals from populations and the selection of individuals with a higher rate of reproduction.

Unlike Darwin's view, this theory rejects the inheritance of acquired characters. It implies that genotype does not depend on phenotype.

Reinforced by the prestige of genetics, the neo-Darwinian interpretation of evolution prevailed. Key arguments in its favor were put forward by the geneticist Dobzhansky and the paleontologist George Simpson.[26] Like genetics, however, neo-Darwinism faced a number of major problems. One of its key postulates was that variations (mutations) were small in scale and could have either positive or negative effects. This was particularly problematic given that nothing was known about the nature of mutations, genes, or gene products, nor about how natural selection acted upon them. It would have been quite legitimate to argue that the "evolutionary synthesis" was just as abstract and divorced from the real world as the formal genetics from which it emerged.

CHAPTER 2

The One Gene–One Enzyme Hypothesis

In 1941, George Beadle and Edward Tatum showed that genes controlled enzyme synthesis[1] and that there was a different gene for each enzyme. This was the first step toward the unification of biochemistry and genetics and was the first major discovery of molecular biology.

To use the term coined by Joshua Lederberg and Harriet Zuckerman,[2] Beadle and Tatum's discovery was "postmature." Given the spectacular developments in biochemistry and genetics that had taken place at the beginning of the century, it is surprising that the discovery was not made earlier.

A number of results, dating back to the origins of genetics, had suggested that there was a connection between genes and the chemical reactions that take place within organisms.[3] The most easily detectable genetic differences are color differences produced by the presence or absence of pigments. By the end of the nineteenth century, many pigments had been chemically characterized, at least in part.

The first precise relation between genes and metabolism was described in 1902 by Archibald Garrod, a physician at Saint Bartholomew's Hospital in London.[4] In 1898, one of Garrod's patients was a young boy suffering from alkaptonuria, a disease in which the patient's urine darkens on contact with air. This disorder was well known, and the chemical responsible for the dark color had been characterized in 1859. It had subsequently been shown to be produced by the transformation of an amino acid—tyrosine—that is present in food.

In 1901, a fifth child suffering from alkaptonuria was born into the same family. By studying his patient and the new sibling, Garrod was able to show that the symptoms were due to a metabolic disorder, the chemical equivalent of an anatomical deformation.

Garrod also noted that the children's parents were first cousins. On studying other examples of children with alkaptonuria, he discovered that in three out of four cases, the children were born to parents who were first cousins. Aware of the work of the English geneticist William Bateson, Garrod concluded that the problem was due to a rare Mendelian factor: "the mating of first cousins gives exactly the conditions most likely to enable a rare, and usually recessive character, to show itself."[5]

In his book *The Inborn Errors of Metabolism* (1909), Garrod described the exact chemical nature of the disorder found in patients with alkaptonuria—in such patients one of the first stages of the metabolism of tyrosine, the splitting of the benzene ring, never occurs. Garrod concluded: "We may further conceive that the splitting of the benzene ring in normal metabolism is the work of a special enzyme, [and] that in congenital alkaptonuria this enzyme is wanting."[6]

Garrod's work was favorably received by English geneticists and biologists, such as J. B. S. Haldane, but it could not lead to a research program because the experimental material—human beings—was not suitable for a genetic and biochemical study. Furthermore, in 1909 genetics was still in its infancy, and population genetics did not even exist. The study of the metabolic pathways involved in the disease also proved problematic; it was not until around 1950 that they were finally unraveled through the use of chromatography and radioactive markers.

Garrod worked in a field that aimed to characterize the chemicals present in organisms—physiological chemistry. This discipline gradually lost ground to biochemistry, which was concerned with fundamental metabolic reactions.

Three other sets of data, collected between the 1920s and the 1940s, also suggested a relation between genes and enzymes. The first was the study of plant pigments, or anthocyanins, carried out by Muriel Wheldale followed by Rose Scott-Moncrieff. The second was Fritz von Wettstein's work on eye color in the butterfly *Ephestia kuhniella.* The third and most important was the work of Boris Ephrussi and George Beadle in Thomas Morgan's laboratory at Caltech and later at the Institut de Biologie Physico-Chimique in Paris on eye color in *Drosophila.*[7]

All these studies met the same obstacles that Garrod had encountered, namely, the complexity of the metabolic pathways and the difficulty of characterizing the substances involved. The subject of these studies was thus not a good choice; Beadle and Tatum's merit was that they studied the genetic control of essential metabolic pathways rather than "superficial" ones, pathways that were well known or at least easy to study experimentally. But the separation of biochemistry and genetics had other causes that flowed directly from the research carried out by the dominant current in genetics, Morgan's school.

Morgan and his students collected a remarkable amount of data on *Drosophila.* They localized several genes to specific chromosomes and studied down to the smallest detail the frequency of different forms (alleles) of the genes and the effect of the position of these genes on the chromosomes.[8] But they ignored two essential problems: the nature of genes and their mechanism of action. Muller was the only one to tackle the first problem. Attempts to solve it led to the formation of a research group (the "phage group") that was to be particularly important in the history of molecular biology (see Chapter 4). The second question was virtually ignored, except by one of Morgan's students, Jack Schultz, who went to work with Torbjörn Caspersson (see pages 140–141). Richard Goldschmidt, a German geneticist and a sharp critic of formal genetics, laughed at those who drew up chromosome maps based on the frequency with which characters were transmitted, while consciously ignoring the link between genes and the characters they control.[9]

Apart from a few rebels, such as C. H. Waddington and Goldschmidt, the geneticists were equally uninterested in the role of genes in development.[10] This apparent lack of interest was even more surprising given that several of the founders of genetics, including Morgan, had initially been embryologists.[11] Stranger still, Morgan's late "conversion" to genetics was linked to the discovery of the role of chromosomes in sex determination.[12]

Some historians have interpreted the geneticists' apparent lack of interest in the very biological problems that had initially oriented them toward genetics as the result of a strategy that was both cognitive and institutional.[13] As noted earlier, genetics could only really develop on condition that it separated, albeit arbitrarily, heredity and gene transmission from development and gene action. From an institutional point of view, this separation led to autonomy for the geneticists, who became isolated from embryologists and from other biologists.

Other, less political and more scientific reasons led Morgan to turn to genetic research. He was convinced that in the future scientists would indeed investigate the mechanism of gene action and the role of genes in development. In the 1930s, however, he felt that the complexity of the relation between gene and phenotype rendered the study of such links premature.[14]

Morgan's caution seems justified, given the problems Garrod encountered in interpreting the chemical defect associated with alkaptonuria, and the difficulties and lack of success of work on anthocyanins and on insect eye pigments. Such caution was all the more tenable considering that molecular biology had only recently begun to make inroads in the field of developmental genetics.

As described earlier, George Beadle had worked with Boris Ephrussi on *Drosophila* eye pigments.[15] This work convinced Beadle that pigment synthesis was the result of a series of reactions, catalyzed by different enzymes that were themselves controlled by genes. But the chemical complexity of this system made it impossible to transform conviction into proof. Because the chemical analysis of the products of gene action was the most difficult part of the study, argued Beadle, the problem had to be looked at the other way around, by starting off from well-known metabolic compounds and studying the genetic control of their production or use.

This required an organism that could be studied genetically, while at the same time permitting a simple biochemical investigation. Beadle chose a fungus, *Neurospora,* that had recently been analyzed genetically by Carl C. Lindegren in Morgan's group. Beadle recruited Edward Tatum, who had experience in the conditions required for growing micro-organisms (culture media and vitamins).

Beadle's question was straightforward. In order to grow, *Neurospora* did not need vitamins (with the exception of biotin) because it could generate these compounds itself. Was it possible to isolate mutants that had lost this ability and therefore needed to be supplied with a given vitamin?

Neurospora has a complicated reproductive cycle: the organism has a haploid form (that is, it contains only one set of chromosomes instead of two, as in most organisms). Each gene is thus present in only one copy—any modification of the gene is immediately expressed because it is not "masked" by the other copy. Two haploids can be crossed to get a diploid form, and the haploid products of this cross can in turn be isolated. Owing to the nature of this reproductive cycle, genetic studies can be carried out in the haploid state, thus facilitating the rapid isolation of mutations.

Beadle and Tatum irradiated *Neurospora* spores in order to produce a large number of mutations. After crossing, they added potentially mutant spores to either a normal medium or to one containing vitamin B1 and vitamin B6. This procedure enabled Beadle and Tatum to isolate several dozen mutant strains that required one or another vitamin to survive. By crossing the strains, they showed that each nutritionally deficient strain was the result of a mutation in a single gene. They then found mutant strains that could not synthesize a given amino acid—tryptophan—and that thus required this substance to be added to the medium in order to grow. They characterized several mutants and genes that enabled them to describe precisely the tryptophan biosynthetic pathway and show that each step was controlled by a different gene.

Beadle and Tatum's results were enthusiastically received. In 1944 Beadle was elected to the National Academy of Sciences, and in 1958, together with Tatum and Lederberg, he was awarded the Nobel Prize. (It is particularly remarkable that during the Second World War, Beadle was able to develop a fundamental research project requiring personnel and materials.) The historian L. E. Kay has uncovered a hidden—or at least forgotten—part of Beadle's work, namely, its applied aspect.[16] Beadle's research had two potential consequences: it could lead to a better understanding of metabolism and perhaps to the discovery of essential metabolic factors, such as vitamins or amino acids. But above all, Beadle and Tatum's mutants provided an indirect but elegant method of measuring the quantity of vitamins or amino acids present in different types of food. The growth of mutant strains of *Neurospora* and hence their weight depended on the supply of amino acids or vitamins.

Beadle and Tatum were well aware of the possibility of applying their research. Tatum had been trained at the University of Wisconsin at Madison, where the Department of Biochemistry had close ties with the dairy, agricultural, and food industries. These links had partly been established by Tatum's father, Arthur Tatum, who was himself a well-known university biochemist.

Beadle and Tatum thus received funding from various pharmaceutical and agricultural companies and from the Nutrition Foundation. This did not prevent Beadle from continuing to receive major financial support from the Rockefeller Foundation, which had backed his project from the outset and had been behind Tatum's recruitment. Most important, the possible applications enabled Beadle to keep his students doing research at a time when the war effort had led to a reduction in the financing of fundamental research through the channeling of money to war-related subjects and the recruitment of young scientists into the army.

It is not easy to measure the impact of Beadle and Tatum's data on the development of molecular biology. As noted, their work was well received. In 1945, Beadle gave the prestigious Harvey Lecture, in which he stated that his work had led to the unification of biochemistry and genetics, which had been separated by "human limitations and the inflexible organization of our institutions of higher learning."[17]

The genetic approach developed by Beadle turned out to be a remarkable tool for the study of metabolic pathways. In the space of a few years, biochemists were able to establish the complex map of organic metabolic pathways, using radioactively labeled molecules and *Escherichia coli (E. coli)*.

Beadle and Tatum's work had indeed led to the experimental association of biochemistry and genetics, but at a more fundamental level, did their study clarify the link between genes and enzymes? The answer to this question is less clear-cut. All of Beadle and Tatum's results, and those obtained by other groups using the same approach, showed that each elementary chemical step in a metabolic pathway is controlled by a single gene. Since each step is under the control of an enzyme, the one gene–one enzyme hypothesis followed logically. But a retrospective, anachronistic reading of this hypothesis is potentially misleading. We now know that the structure of each protein and thus of each enzyme is "coded" by a separate gene. But the notion of coding, of the genetic determination of the detailed structure of proteins and enzymes, was completely absent from Beadle's thinking.

Two points will help to scale down the importance of this discovery, or at least to place it in its proper scientific context:

- The fact that a single gene seemed to control each metabolic enzyme did not mean that one gene was sufficient for the synthesis of that enzyme. Enzymes are proteins; because proteins are formed of numerous amino acids, it was thought that for a given enzyme to be synthesized, several enzymes and thus several genes would be necessary. The gene revealed by Beadle and Tatum's experiments was the one that conferred the final structure, form, and specificity on the newly synthesized enzyme. Delbrück, the leader of the "phage group" and one of the acknowledged founders of molecular biology, was skeptical of Beadle and Tatum's results.[18] Like most biochemists, Delbrück thought that several genes controlled the synthesis of a protein or enzyme. Some of these genes—including those discovered by Beadle and Tatum—were specific to a single enzyme. Others controlled the synthesis of several enzymes and therefore had not been selected in Beadle and Tatum's procedure. This implied that the one gene–one enzyme hypothesis was the consequence of an experimental artifact.

- On the other hand, Beadle and Tatum's results seemed to bring genes and enzymes closer, even to the extent of identifying them with each other. Many biologists had an "enzymatic" view of genes and reasoned as follows:[19] the astonishing catalytic properties of enzymes were responsible for organic specificity; genes control both the organism and enzyme synthesis—thus the simplest hypothesis was that genes were self-synthesizing, self-replicating enzymes. This protein or enzymatic theory of genes was based on results from experiments on bacteriophage reproduction carried out by John Northrop's group at the Rockefeller Institute, and Wendell Stanley's work on the tobacco mosaic virus, or TMV (see Chapter 6).

The key point is that the immediate interpretation of Beadle and Tatum's experiments was not that retained by history. This will be particularly important when it comes to understanding the context in which Oswald Avery discovered the chemical nature of the transforming factor in 1944 (see Chapter 3).

In the provocatively titled book *Where the Truth Lies,* Jan Sapp has suggested that Beadle and Tatum's results were anticipated by the work of the German geneticist Franz Moewus, who was thus the true founder of genetic biochemistry.[20] Sapp argues that the importance of Moewus's research was not recognized because accusations of fraud discredited his ideas and work.

In the 1930s, Moewus was one of the first biologists to attempt a genetic study of a micro-organism. He chose to study an alga, *Chlamydomonas,* and described the principal phases of its reproductive cycle. Together with the German biochemist Richard Kuhn, Moewus showed that the various stages of the *Chlamydomonas* reproductive cycle were controlled by carotenoid hormones that were synthesized and converted through the action of enzymes, each of which was controlled by a different gene that Moewus localized on the alga's chromosomes.

At first, Moewus's results were well received, although some doubts were expressed. From a statistical point of view, some of the data were simply too good to be true, and other scientists could not get hold of the strains that Moewus had worked with. Finally, his experiments could not be replicated, even in his presence.

There can be no doubt that Moewus was the first to have seen the relevance of micro-organisms for forging links between biochemistry and genetics. But his work did not represent a break with earlier studies carried out on other systems. It was analogous to Beadle and Ephrussi's research on eye color in *Drosophila.* Unlike Beadle and Tatum, Moewus did not approach the problem from the other end—studying well-known biochemical reactions by genetic means. Beadle and Tatum's great merit was that they obtained enough observations for the one gene–one enzyme hypothesis to take on a quantitative and not merely a qualitative meaning.

Despite the interesting material it deals with and its thought-provoking analyses, Jan Sapp's book is unsatisfying. It is a caricature of some contemporary trends in the history of science. Sapp does not really address the question of whether or not Moewus manipulated his results. For Sapp, even if the accusation were true, it would simply mean that Moewus was no different from other scientists, all of whom "correct" their experimental data. This *a priori* judgment prevents Sapp from dealing with a number of questions relating to the validity of Moewus's results, such as whether, half a century later, the various hormones characterized by Moewus have been shown to exist, and if so, whether they have the structure and function that Moewus attributed to them.

It is, of course, true that anyone who asks these kinds of questions must be hopelessly blinkered by a "realist" view of science! For Sapp, "discoveries do not exist independently of their inventors," and "when Moewus was excluded from this scientific domain, his discoveries vanished with him."[21] There we have it: if science has no relation to reality, why bother giving a full and precise description of the facts?[22]

CHAPTER 3

The Chemical Nature of the Gene

In retrospect, the first experiment that convincingly showed that genes are made of DNA was carried out by the American Oswald T. Avery and his associates, and published in 1944 in the *Journal of Experimental Medicine.*[1] Yet the earliest descriptions of the origins and development of molecular biology did not mention this study. Another experiment, performed eight years later, produced the same result[2] and is often presented together with or instead of Avery's experiment. This has given rise to a substantial literature according to which Avery was an unrecognized genius, molecular biology's equivalent of Mendel.[3]

The truth is both simpler and more complex—simpler, because Avery was not unknown; more complex, because his discovery was not understood nor its importance recognized at the time. In the words of H. V. Wyatt, everybody knew about Avery's discovery, but the information did not "become knowledge."[4]

Avery worked in the prestigious Rockefeller Institute in New York, where he spent most of his scientific career studying pneumococci (pneumonia-causing bacteria).[5] After infection, the organism reacts by producing antibodies against the pathogenic agent. This had led to the characterization of various kinds of pneumococci. Antibodies react with the bacterial envelope, or capsule, which is made up of sugars. Avery had made an important contribution to the characterization of the structure of this capsule in different types of pneumococci and had acquired a reputation for the high quality of his work.

Pneumococcal infection can be produced in mice. Because it caused such serious diseases, the pneumococcus became a model for the study of human infectious disease. Avery's work thus fell squarely within the scope of the Rockefeller Institute's research interests: understanding pathological phenomena at a fundamental level using the most up-to-date techniques from physics and chemistry.

Under certain culture conditions, pneumococci spontaneously shed their sugar capsules. These R ("rough") pneumococci are not infectious, unlike S ("smooth") pneumococci, in which the capsule is intact, thus making the bacterial colonies appear smooth. The sugar capsule inhibits phagocytosis of the pneumococci by the organism's white blood cells, thus accounting for the bacterium's pathogenic power. When antibodies attach themselves to the capsule, they encourage phagocytosis and thus help defend the organism.

In 1928, the British physician Fred Griffith discovered the strange phenomenon of transformation. When a mouse was simultaneously injected with nonpathogenic R pneumococci (derived, for example, from type II bacteria) and type III S pneumococci killed by heat, the animal rapidly succumbed to infection and died. Griffith was able to isolate colonies of the virulent type III S form from the blood of the dead mouse. The only possible interpretation was that the type III pneumococci—virulent but dead—had transformed the nonpathogenic live type II pneumococci.

In 1931, Martin Dawson and Richard Sia reproduced this result *in vitro,* while J. Lionel Alloway showed that it was possible to transform *in vitro* nonvirulent pneumococci of one type with extracts of killed virulent pneumococci of another type. Alloway's experiment opened the way to the purification of what became known as the transforming factor—whatever it was in the extracts of killed pneumococci that was responsible for the transformation of pneumococcal type.

In 1935, Avery began his attempts to purify the transforming factor. Together with Colin MacLeod and Maclyn McCarty, Avery continued his work—with a few interruptions—for nearly a decade.[6] Their 1944 article, published in the *Journal of Experimental Medicine,* represented a substantial amount of work. Several pages were devoted to explaining well-defined and easily reproducible conditions for carrying out transformation *in vitro,*[7] and the purification of the transforming factor was described in some detail. But the key part of the paper was the characterization of the purified transforming factor via a series of physical and chemical methods.

All these techniques pointed in the same direction: the transforming factor was not a protein but a deoxyribonucleic acid. The factor was able to withstand temperatures that would denature a protein. Colorimetric tests showed that the purified transforming factor contained only DNA, with no traces of protein or of ribonucleic acid (RNA). Simple chemical analysis confirmed that proteins made up less than 1 percent of the purified material. Enzymatic tests showed that the transforming factor was not affected by enzymes that cleaved proteins, nor by phosphatases that degraded RNA, but that it could be destroyed by unheated serum. This last point was important: serum was known to contain an enzyme capable of degrading DNA.

Using the state-of-the-art techniques from physics and chemistry that were the hallmark of the Rockefeller Institute (see Chapter 9), Avery was able to show that the purified transforming factor had a high molecular weight, and was probably a nucleic acid.

Finally, a significant part of the paper was devoted to an immunological analysis showing that the purified material did not react with antibodies directed against the capsule. This seemed to suggest that although the transforming factor was responsible for the formation of the capsule, it did not itself contain any sugars: it was therefore chemically different from the chemical structure it transformed. The paper concluded with a discussion of the nature and mechanism of action of the transforming factor.

The usual explanation for the fact that Avery's discovery did not have a significant impact is that his paper was not read, or at least not read by those who would really have understood it, and in particular it was not read by geneticists. The *Journal of Experimental Medicine* was aimed more at physiologists and pathologists than at protein biochemists or geneticists. Furthermore, the title and the abstract of the article, which were essentially centered on transformation, did not underline the importance of the final result. Even the discussion was cautious—Avery did indeed put forward the hypothesis that the transforming agent might be related to or of the same nature as genes or viruses (especially the tumor viruses discovered some years before by Peyton Rous), but he merely outlined a possible relation without going any further. This caution was symptomatic of Avery's personality. He was sixty-seven years old at the time, and preferred bench work to research administration. He never sought an important position and rarely attended conferences.[8]

None of these limitations, however, prevented his paper from being read and discussed, as H. V. Wyatt has clearly shown.[9] It should be remembered that the Rockefeller Institute was one of the most prestigious research institutes in the United States, that Avery was well known and his work widely appreciated, and that the journal in which he published his article had a large circulation and a solid reputation.

Avery's work did not go unnoticed; his discovery, popularized by journals like *American Scientist,* was brought to the attention of the key biochemists and geneticists and the main practitioners of the nascent science of molecular biology. The result seemed to be very important, but it raised so many problems that it was difficult to appreciate its real significance.

With characteristic bluntness, Max Delbrück said of Avery's result, "You really did not know what to do with it."[10] This comment reveals the state of knowledge at the time and the real status of the models that were often used as a substitute for understanding.

Transformation was a new and unusual phenomenon that at the time appeared to be limited to pneumococci. Even its extension to the bacterium *E. coli* by André Boivin in France did not support Avery's view—these results could not be reproduced by other laboratories, which meant that transformation became even more marginalized.[11] Furthermore, the pneumococcus was poorly understood in terms of both its make-up and its biochemical nature. Prior to Avery's work, the only nucleic acid that had been characterized in this bacterium was RNA. The existence of genes in bacteria was not universally accepted[12]—some scientists such as the British biochemist Sir Cyril Hinshelwood thought that the properties of bacteria, including adaptation and mutation, could be explained by changes in biochemical equilibria (see Chapter 5).

Another limitation was the fact that nucleic acids were poorly understood at the time. The chemical structure of nucleotides and the nature of the bonds that bind them in polymers were still matters of dispute. The dominant notion was the tetranucleotide model proposed by Phoebus A. Levene in 1933, according to which DNA was formed by the repetition of an elementary unit—a chain of four bases (adenine, thymine, guanine, and cytosine). For Levene, therefore, the DNA molecule was monotonous. Its function was poorly understood, although detailed cytological experiments mainly carried out in Sweden by Torbjörn Caspersson had shown that it was associated with chromosomes. According to the nucleoprotein chromosome model, DNA was the material support that carried the proteins thought to be responsible for genetic specificity.

A few years earlier, the American biochemist Fritz Lipmann had discovered that the nucleotide adenosine triphosphate (ATP) acted as an intracellular energy reserve. This had suggested an energy-related dimension to the nucleoprotein units that were known to exist in chromosomes—perhaps DNA played a role in providing the energy needed for gene replication? Avery's discovery thus took place in a context that was extremely unfavorable to the idea that DNA played a specific role. Avery himself was well aware of this problem; one of the questions discussed in detail in his paper was the fact that the DNA molecule appeared to be nonspecific.

In addition to these difficulties, which were caused by ignorance of the structure and properties of nucleic acids, another problem with this study was that it came up against the widely held view (which Avery agreed with) that proteins were the most likely candidates for genetic specificity[13] (see Chapter 1). The idea that proteins were the essential constituent of genes was not new: it was accepted by both geneticists like Muller and the young molecular biologists of the "phage group" whose work will be described in the next chapter.[14] W. M. Stanley gave support to this idea when he isolated a plant virus—the tobacco mosaic virus (TMV)—as a pure protein (see Chapter 6). This protein model of genes agreed with all the experimental data that had been accumulated over the previous half-century, and which showed that proteins were the "carriers" of organic specificity. This specificity was first revealed by the wide range of cellular reactions controlled by proteins. John Northrop and John B. Sumner, two of Avery's colleagues at the Rockefeller Institute, had purified and crystallized several enzymes and had confirmed that they were proteins.

Specificity of protein form and function can be studied by making antibodies against proteins. Immunological reactions were originally used to reveal structural differences between protein molecules that no physicochemical method could detect. Immunological criteria later became "the" specificity criteria—a molecule was considered "specific" if it could induce specific antibodies. Up until this point, DNA had failed to pass the test.

Other arguments supported the hypothesis that proteins were the main constituents of genes: they were definitely one of the two key components of chromosomes, and Beadle and Tatum had shown that enzymes responsible for various metabolic reactions were closely controlled by genes. Beadle and Tatum's work tended to link genes and enzymes and, as far as many biologists were concerned, had reinforced the more or less conscious identification of genes with enzymes and proteins.

Furthermore, Avery had obtained a relatively large quantity of the transforming factor: the trace levels of contaminating proteins that were found in it could explain the observed data. Or so the partisans of the protein theory of genes, such as Alfred E. Mirsky of the Rockefeller Institute, believed.

In the years that followed, Avery's colleagues improved the purification of the transforming factor. Contamination by amino acids—constituents of proteins—was reduced to less than two parts in ten thousand. The presence of these trace amino acids was in fact not due to protein contamination but to the degradation of a nucleotide (adenine). The identification of DNA as the transforming agent was confirmed by its direct inactivation by deoxyribonuclease, an enzyme recently isolated at the Rockefeller Institute. Nevertheless, these confirmations and improvements of Avery's experiment did not shake the convictions of those who thought that proteins played an essential role in heredity.

The main problem preventing the importance of Avery's discovery from being recognized was not so much the difficulties posed by DNA or the objections raised by the supporters of the protein theory of genes, as the impossibility of interpreting the data. Avery's letters to his brother and the testimony of his colleagues show that despite the cautious conclusion to his 1944 paper, he was in fact profoundly convinced that the isolated material—the transforming factor—was a pure gene.[15]

Although pneumococcal transformation had been an ideal system for discovering the chemical nature of genes, it was most certainly not suited to the extension and generalization of this result, and even less so to its interpretation.[16] Transformation was a highly particular phenomenon, mainly of medical interest, and was limited to pneumococci. In subsequent years, Avery's colleagues showed that transformation could also apply to other characters apart from the capsule. But extending the phenomenon to other bacteria, for example, to *E. coli,* turned out to be difficult. Furthermore, the observation of transformation in pneumococci is particularly laborious and difficult; indeed, Avery's paper emphasizes the care needed in carrying out experimental observations of transformation. The most intriguing aspect of the result was how exactly a nucleic acid could control the structure of a capsule made of sugars. The problem was well presented in Avery's 1943 Rockefeller Institute report: "The transforming principle—a nucleic acid—and the end type of synthesis it evokes—the type III polysaccharide—are each chemically distinct . . . The former has been linked to a gene, the latter to a gene product, the accession of which is mediated through enzymatic synthesis."[17]

Avery's model required a stage involving the presence of an enzyme or a protein. He had to find out how the nucleic acid—which was thought to be a nonspecific compound—could control protein activity, or more precisely, control the enzymes responsible for capsule synthesis. The supporters of the protein theory of genes did not have this problem. It was easier for them to explain how proteins (genes) could control the activity of other proteins than how nucleic acid could control enzymes. Or at least, so it seemed.

Two obstacles had to be overcome before it could seriously be considered that genes were formed of DNA. Contrary to what might be expected, the idea of a genetic code was not a precondition. This idea was in fact introduced gradually, much later on, after James Watson and Francis Crick's discovery of the double helix structure of DNA in 1953. In fact, it was only really accepted in 1961, when the genetic code was deciphered. The first necessary step was the separation of the dual problems of the nature of the gene and its mechanism of action.

Unlike biochemists, geneticists had long been prepared for this kind of approach to the problem. And even more than the geneticists, physicists saw genetics as information transfer, and the gene as a carrier of information.[18] This insight in fact resulted from their lack of understanding and their despair when faced with the seemingly irrational complexity of biochemistry. The physicists were able to approach the problem from another angle and abstract the notion of specificity from its biochemical and protein connotations. They separated the gene from its effects within the cell and made it possible to conceive of relations between genes and proteins apart from a direct protein-protein interaction.

This alternative way of looking at genes could prevail only if it was associated with a new experimental system in which the study of the transfer and replication of genes was more important than the characterization of their functions. This system could have been the study of plant or animal viruses; instead, it turned out to be the bacteriophage, an entirely novel experimental system described in the next chapter.

This alternative conception of the role of genes gradually led to the abandonment of the idea that genes, like catalysts, control organic functions from a distance, and its replacement by the current model of molecular biology, in which genes determine the development and functioning of the organism down to the smallest chemical detail.

Avery's experiment was difficult for his contemporaries to accept, and even more difficult for them to interpret. Although the experiment was well known, it did not provoke the conceptual revolution that might have been expected.

Avery was perhaps not as single-minded in his research as he could have been. Between 1937 and 1940, together with some of his colleagues, he turned away from transformation for a while.[19] Nicholas Russell has raised the possibility that the discovery of sulfamides (which made Avery's earlier research on the development of a pneumococcal vaccine superfluous) led him to focus his final research years on the transforming factor.[20] By training and personal temperament, Avery was more suited to solving precise problems than to defining long-term research programs.

It should not be thought, however, that Avery's results had no influence on the subsequent development of molecular biology. The fact that his work was incompatible with Levene's tetranucleotide model intrigued Erwin Chargaff, an Austrian-born American biochemist. Chargaff repeated the experimental measurements on which Levene's model was based, and accurately measured the percentages of the four nucleotides in the DNA of several different species.[21]

Chargaff's data were significant in a number of ways. He showed that the proportions of the four nucleotides changed depending on the species being studied, thus removing a major obstacle in understanding the genetic role of DNA. Furthermore, Chargaff separated the nucleotides by a highly sensitive chromatographic technique, which was subsequently widely applied in molecular biology. Last but not least, he almost reluctantly found a certain regularity in the proportions of the four nucleotides and the four bases that constituted them. In all the DNA samples he studied, he found that the proportions of adenine and guanine were equal to those of thymine and cytosine respectively. This result came to be known as Chargaff's rule, and, as will be seen, it played an important part in the discovery of the double helix structure of DNA.

H. V. Wyatt has shown that the "phage group" was aware of the importance of Avery's result, and that some members of the group tried to reproduce it. For others, it simply meant they had to take more seriously the hypothesis that DNA was the essential component of bacteriophage reproduction.[22]

Finally, and perhaps most important, Avery's experiment was the first to undermine the conviction of many biochemists and geneticists that genes were proteins. Although it did not convince them that genes were made of DNA, it opened the door to that possibility, revealing a scientific trail that would be blazed in the years to come.

CHAPTER 4

The "Phage Group"

The term "phage group" refers to all those scientists who, between 1940 and 1960, used bacterial viruses ("bacteriophages") as a model system to study how organisms function. The group was never an organized structure, nor did it have any official existence as such. The research carried out by its members was extremely diverse and evolved over the years. What gave the group its identity was a state of mind, a new approach to biological problems. This was largely the result of the influence of Max Delbrück,[1] who is universally acknowledged as the founder and driving force behind the group.

Delbrück was born into an upper-class German family in 1906. His father was a professor of history at Berlin University. In 1930, Delbrück was awarded his doctorate in theoretical physics.[2] The following year, funded by a grant from the Rockefeller Foundation, he joined Niels Bohr's laboratory in Copenhagen. His meeting with Bohr, and especially Bohr's lecture on "Light and Life"[3] (delivered in Copenhagen on August 15, 1932), had a decisive influence on Delbrück and led him to turn to biology. (Bohr's lecture will be discussed further below, together with his ideas about biology.) Bohr was not a very clear speaker, and the lesson that Delbrück took home from the lecture was probably not what Bohr had intended. What Delbrück understood was that even if the study of biological molecules were to be taken as far as possible, it would still not lead to an understanding of life. Life could be understood only through an entirely new approach, which would complement existing methods. On his return to Berlin, Delbrück set up a small informal group of biologists and physicists, which met at his mother's house. His first piece of biological research was carried out in collaboration with the Russian geneticist Nikolaï Timofeeff-Ressovsky and the German physicist Karl Zimmer.[4]

The idea behind this study was taken from nuclear physics. Physicists could not study the nucleus directly; they had to bombard it with particles. By varying the size and energy of these particles and comparing the variation with the results of the bombardment, it was possible to deduce some of the properties of the target—in this case, the nucleus.

Delbrück decided to use the same approach to study the gene. In 1927, Muller had shown that mutations could be induced by X-rays. Timofeeff-Ressovsky's job was to determine the number of mutations induced in *Drosophila* by irradiation with different energies delivered at different temperatures. Zimmer converted radiation doses into the number of ion pairs formed (it was assumed that ion pairs were responsible for the mutagenic effect). On the basis of these experiments, Delbrück estimated the size of the gene to be a few thousandths of a millimeter. Variations in mutation rate as a function of temperature were interpreted in terms of a quantum model of the gene, according to which the gene had several stable energy states, like a molecule; a mutation was interpreted as a passage from one stable state to another.

Their paper (known as the "Three-Man Paper") was published in 1935 in German, in a Göttingen-based journal with a small circulation.[5] Nevertheless, it came to the attention of Schrödinger, who devoted considerable space to it in his book *What Is Life?*[6] Salvador Luria, a physicist working in Enrico Fermi's laboratory in Rome, and who later worked closely with Delbrück, also heard of the paper.[7] On the basis of this study Delbrück and Timofeeff-Ressovsky subsequently published an article in *Nature* concluding that cosmic radiation had little effect on speciation.[8]

Several research groups had used, or would use, similar approaches to study gene structure and the nature of mutations, but in the absence of any data on the chemical nature of genes, the results were difficult to interpret. Although Delbrück's model of mutation by ionization turned out to be wrong, it was nonetheless a "successful failure."[9] These and similar experiments helped to show that genes could be studied with the tools of physics.

In 1937, again with a grant from the Rockefeller Foundation, Delbrück visited the most important U.S. genetics laboratories. On October 15, 1937, he arrived at Thomas H. Morgan's laboratory at Caltech, in Pasadena. Delbrück was disappointed with *Drosophila* as a model system—he thought it was too complex to reveal the secret of life. Ever since his meeting with Bohr, he had been convinced that to discover this secret, only the simplest biological system would do. The principles of quantum mechanics had been discovered only when matter had been studied at its most elementary level—that of the atom and its constituents.[10] The same principle should operate in biology. Contradictory experiments and paradoxes—which for a physicist are the starting point for new theories—appear only when simple systems are studied.[11] For Delbrück, this elementary biological system was the bacteriophage, which he discovered during his visit to Morgan's laboratory. Emory Ellis, who had recently started studying phage, worked in the same Caltech department as Morgan.[12]

The bacteriophage, or bacterial virus, immediately struck Delbrück as being particularly suited to the study of the key characteristic of life: self-replication.[13] Bacteriophages infect bacteria and multiply rapidly inside their hosts. They appeared to be elementary biological particles—they were less than one ten-thousandth of a millimeter in length, which meant that, according to Delbrück's earlier study, they were smaller than genes.

Bacteriophages had already been widely studied, first by the man who discovered them in 1917—the Canadian-born French scientist Felix d'Hérelle—and, at the same time, by the British bacteriologist Frederick Twort, then by several other groups, particularly at the Pasteur Institutes in Paris and Brussels, and at the Rockefeller Institute in New York.[14] Despite their small size, phages could be easily detected: a layer of bacteria was grown on the surface of a Petri dish; following infection with the phage, light plaques appeared ("lysis plaques") where the bacteria had been killed and lysed by the bacteriophage.

Nevertheless, the nature of phage and its relation to bacteria remained unclear and was widely debated.[15] Some scientists thought that the bacteriophage was a real infectious particle; others, such as Albert Krueger and John Northrop, saw phage multiplication as evidence of nothing more than the autocatalytic transformation of an inactive protein precursor that was present in bacterial cells before the addition of phage. According to this widely accepted model, bacteriophage replication—which showed an s-shaped ("sigmoid") kinetic—was analogous to the autocatalytic activation of enzymes such as trypsin.[16]

Delbrück began to work with Ellis, who was already studying the phage growth curve. The first step was to develop statistical tests that could validate their method of counting phages. They showed that bacteriophage multiplication was irregular, entailing a series of steps and plateaux. They interpreted the sudden increases in the number of phages that took place every thirty minutes at 37°C as the result of bacterial lysis and the liberation of tens or hundreds of bacteriophages. The phages rapidly attached themselves to new hosts, which they infected, and thirty minutes later the sudden increase would be seen again.

The rapidity with which the bacteriophages interacted with the bacteria, and the existence of a discontinuous growth curve, were both novel observations. This experiment, known as the single-stage growth experiment, was published in the *Journal of General Physiology*, and had a major impact.[17] The experiment was striking because of both its clarity and its elegant statistical methods. But what was most important was that it was in complete opposition to the autocatalytic model proposed by Krueger and Northrop. There was no sign of any sigmoid curves in Ellis and Delbrück's results.

Delbrück's Rockefeller Foundation grant was renewed in 1938. Because of political developments in Germany, he wanted to remain in the United States. At the end of 1939 Vanderbilt University, in Nashville, Tennessee, offered him a job teaching physics. This permanent position enabled Delbrück to carry out his work on phage in the years that followed.

At this point, it is difficult to speak of a "phage group." Ellis, Delbrück's only collaborator, had to stop work on phage and return to research more directly centered on cancer. The phage group emerged as a result of Delbrück's meeting with Luria at the end of 1940 and their first joint research, carried out in the summer of 1941 at Cold Spring Harbor on Long Island.[18] This collaboration between Luria and Delbrück also gave rise to bacterial genetics (Luria's scientific career and his first experiment with Delbrück are described in Chapter 5). The third founding member of the group, Alfred Hershey, joined Luria and Delbrück in 1943.

The members of the phage group carried out many experiments as they tried to understand bacteriophage replication. One of Luria's experiments showed that if the same bacterium is infected with two different kinds of phage, a kind of interference or mutual exclusion takes place between the two phages. A series of other experiments was carried out in order to uncover what happened to the bacteriophage during the "eclipse phase," its period of replication.

Any attempt to analyze precisely the results obtained by the phage group between 1940 and 1953 (that is, until the discovery of the double helix structure of DNA) reveals a striking contrast between the group's reputation and the relative scarcity of its results. Many biologists will be surprised by this idea; after all, the phage group was responsible for the birth of bacterial genetics, while an experiment conducted by Alfred Hersey and Martha Chase (see below) showed that DNA was the hereditary material. But the link between these two discoveries and the phage group is, in fact, quite complex. Luria and Delbrück's 1943 experiment, which showed that there were mutations in bacteria, was in reality a marginal by-product of their work on phage. Soon afterwards, Delbrück turned away from bacterial genetics and concentrated on phage.[19] And as will be seen, Hershey and Chase's discovery was in fact merely a new proof of the results found by Avery and his coworkers. It was mainly because of the phage group's prestige that the Hershey-Chase experiment had such an impact.

But results from the key theme of the phage group's work—bacteriophage replication—were relatively unimportant. In particular, Delbrück's initial dream of understanding replication as a simple phenomenon without opening the biochemical "black box" proved to be an illusion.

In 1942, Thomas Anderson used electron microscopy to study the bacteriophage[20] and showed that it was not an elementary biological particle but a complex organism, highly organized and structured (similar images had been obtained in 1941 by Ernst Ruska in Germany[21]). Following this discovery, experiments that disturbed bacteriophage replication through various kinds of physical treatment, in particular by radiation, no longer appeared appropriate as a way of characterizing the various stages of replication.

At this point, Luria's understanding of the problem was as follows.[22] The physicists' way of investigating the gene was like someone shooting at a tree with guns of different calibers, in order to find out about the tree's fruit. Depending on the effectiveness of the different sizes of bullet, one could deduce whether the fruit was the size of a cherry or the size of an apple. But to make any further progress it would no longer be sufficient to shoot at the tree: the fruits would have to be isolated and characterized.

The phage group's initial research project was thus a failure. Not only had no new physical principle been discovered and no paradoxical result been recorded, but research on bacteriophage was becoming increasingly influenced by biochemistry,[23] which Delbrück found depressingly complex.

The scarcity of the results obtained by the phage group seems to have been inversely proportional to its fame. There are several reasons for this apparent paradox. Membership of the group was not restricted to people who worked with Delbrück. Between 1940 and 1945, many laboratories began to work on bacteriophage.[24] For example, Milislav Demerec, a Cold Spring Harbor *Drosophila* geneticist, started to study bacteriophage, while Leo Szilard, a physicist who had helped set up the Manhattan Project, organized monthly discussion meetings on phage. A large number of scientists, trained in both biology and physics, were thus beginning to study the phage.

This rapid growth of the phage group was partly the result of Delbrück's personal influence,[25] which can be explained in a number of ways.

Above all, his methods were extremely elegant. Using simple statistical tests, he was able to clarify confused questions. His approach also benefited from the prestige that surrounded any technique or concept that had its origin in physics. But most important, Delbrück showed that it was possible to develop a revolutionary approach to biology. The new research dealt with what was specific about life: self-replication. The credo of Delbrück and his colleagues, as set out later, was that the same principles should be able to explain the functioning and reproduction of all organisms, from the virus to mankind.

The allure of these ideas was reinforced by Delbrück's exceptional charm: he had a brilliant mind and a charismatic personality. He was both kind and somewhat brutal; at the end of a seminar Delbrück would often say to the speaker: "This was the worst seminar I ever heard."[26]

On its own, however, Delbrück's charisma would not have made the phage group famous. Delbrück also introduced a new style into biological research. In his youth he had been profoundly influenced by the atmosphere in Bohr's group in Copenhagen, with its apparent absence of hierarchies, freedom of discussion, and a close mixture of work and pleasure (for example, Bohr's young physicists would often go on hiking trips in the mountains together). Delbrück introduced this new research style into the phage group, first in his Nashville laboratory, then, beginning in 1947, at Caltech. In biology more than in other disciplines, this marked a significant change in the way laboratories functioned. Delbrück wanted his researchers to spend at least one day a week away from the bench, simply thinking about their experiments. Experiments and papers were discussed by the whole group, and were subject to lively criticism. These discussions (which often took place during camping trips to the California desert) and the many parties that were hosted at the Delbrück home made a great impression on the participants. A visit to Delbrück's laboratory was a must for anyone starting out in the new science of molecular biology.

In addition to this "psychological" influence, there was the more direct role of the annual Cold Spring Harbor practical course on bacteriophage, the first of which took place from July 23 to August 11, 1945. A certain level of mathematical ability was required, because the course was essentially aimed at physicists who wanted to do biology. Because Delbrück had become famous through Schrödinger's presentation of his work in *What Is Life?* any physicist who wanted to turn to biology was obliged to attend the Cold Spring Harbor course. The course created an informal network centered on Delbrück and the founders of the phage group. Furthermore, participants all learned to use the same material and the same techniques, thereby reinforcing the influence of Delbrück's approach.[27]

There are at least two types of scientific experiment, each with a different nature and function. Avery's experiment (discussed earlier) is an example of the first type: without any *a priori* expectations, a scientist discovers a new and unexpected phenomenon. The experiment carried out by Hershey and Chase in 1952 is an example of the second type: the result may already be known, and the aim of the experiment is simply to demonstrate it clearly.

The Hershey-Chase experiment was in fact made up of a series of experiments, all of which tended to show that bacteriophage DNA is fundamental for phage replication.[28] History has focused on one of these experiments, the "mixer" or "Waring blender" experiment. This was one of the first experiments in molecular biology to use radioactively labeled molecules. The bacteriophages were labeled either with sulfur (^{35}S), which was incorporated into proteins, or with phosphorus (^{32}P), which was incorporated into DNA. The radioactive phages were added to a culture containing bacteria and the mixture was stirred in the famous blender. This vigorous agitation removed 80 percent of the phage proteins from the bacterial surface, but only 30 percent of the phage DNA. Furthermore, it did not prevent either the bacteria from being infected or the phage from growing. (We now know that when the phage lands on the bacterial surface, it quickly injects its DNA into it.)

Experiments described in the same 1952 article in which the results of the "mixer" experiment were published showed that the phage was merely a protein shell containing (and protecting) the inner DNA: a drastic reduction in salt concentration could separate the proteins from the DNA, which could then be degraded by enzymes. Other experiments showed that DNA was released from the phage during its interaction with the bacterial membrane: if the phage was adsorbed onto killed bacteria or membrane fragments, DNA became available for enzymatic digestion. Finally, the offspring of radioactive phage contained less than 1 percent of the original protein (less than 1 percent of the radioactive sulfur), but at least 30 percent of the DNA (30 percent of the ^{32}P).

All the experiments described in the article thus showed that the part of the bacteriophage responsible for replication was DNA. The protein shell that enclosed the phage DNA was simply there to protect it and inject it into the bacterium like a syringe.

Early histories of molecular biology attributed the discovery of the genetic role of DNA to Hershey and Chase.[29] It took protests from Avery's colleagues for his experiment to be restored to the pantheon of molecular biology. Even today, molecular biology textbooks often put the two experiments on an equal footing, ignoring the fact that they took place eight years apart.

H. V. Wyatt has shown that many books describing the Hershey-Chase experiments "blend" or distort different aspects in order to make them more convincing;[30] for example, the percentages of sulfur and phosphorus that remained associated with the bacteria following agitation are often "massaged" in the "right" direction. In other books, it is suggested that the two kinds of label—sulfur for the proteins and phosphorus for the DNA—were added in the same experiment. In still other accounts, the additional experiments described in the 1952 paper are not even mentioned.

These omissions and alterations are serious because they make it impossible to understand why Hershey and Chase's results were so readily accepted. They were in fact the end-point of a long series of convergent experiments. The first electron microscope images showing phages attached to the bacterial surface but not penetrating the cell implied that only a part of the phage was responsible for its reproduction. The shape of the phage—a kind of syringe with paws—suggested that it injected one of its constituents into the bacterium. The phage was chemically very simple, consisting only of DNA and proteins. Immunological experiments had shown that the proteins formed the phage envelope, whereas DNA seemed to be found within the protein shell.

Avery's result had gradually become an accepted part of scientific thinking.[31] All the members of the phage group knew of Avery's work, and it had encouraged them to study the DNA found in the phage. Furthermore, Levene's tetranucleotide model had been superseded by the work of Chargaff and others, thus removing a series of obstacles to the idea that DNA had a specific function in self-replication. Biochemical studies by André Boivin with Roger and Colette Vendrely, together with cytochemical data, had shown that DNA was present in the nucleus in constant amounts,[32] except in sperm, where there was only half as much.[33] All these experiments helped to make it possible to conceive of DNA as having an essential role in genetic phenomena.

Hershey and Chase's results were thus rapidly accepted. Presented for the first time in Paris, they were outlined for a second time in July 1953 at the Cold Spring Harbor conference, at the same time as the double helix structure of DNA was presented by Watson and Crick (see Chapter 11). In people's minds, the two sets of results became associated and were considered additional proofs of the genetic role of DNA.

Nevertheless, it is impossible not to compare Hershey and Chase's experiment with Avery's. Mirsky of the Rockefeller Institute had criticized Avery's results—or rather, their interpretation—because there were traces of proteins (0.02 percent) in the DNA that could conceivably account for the transforming power of the pneumococci. But in the Hershey-Chase experiment, at least 1 percent of bacteriophage proteins penetrated the bacteria, as did unlabeled proteins with no sulphur-containing amino acids, all of which might nevertheless have contributed to phage replication. Despite this problem, the results rapidly convinced the entire biological community. This disparity shows that a scientific experiment does not have intrinsic value: it counts only to the extent that it forms part of a theoretical, experimental, and social framework.

Things had changed between 1944 and 1952. Everyone now thought that DNA might play a role in heredity. Avery had found an unexpected result, and had gone on to answer a question nobody was asking: What was the nature of the hereditary material? Radioactive labeling made it possible to choose between one of the two components that potentially played this role.

The two experiments were thus carried out in very different conceptual environments. Furthermore, Avery was an isolated researcher who did little to draw attention to his data, whereas Hershey and Chase's results were spread by the informal but highly efficient phage group network. Hershey and Chase's experiment also benefited from the publicity given to Watson and Crick's virtually simultaneous discovery of the double helix structure of DNA.

CHAPTER 5

The Birth of Bacterial Genetics

Microbes were first discovered by Anton Van Leeuwenhoek in the seventeenth century, but it was not until the work of Pasteur, at the end of the nineteenth century, that scientists began to understand how they reproduce. They finally came to accept that these organisms are not the result of spontaneous generation[1] but, like primitive plants, reproduce by division.

Genetics had bypassed bacteria, focusing on organisms with hereditary variations that were easy to observe. Bacteria, which appeared not to reproduce sexually, seemed to have very complex life cycles that enabled them to go through different states or forms. The transformation of S pneumococci into R pneumococci was thus interpreted in the light of "cyclogenic" theories.[2] This confused understanding of bacteria blocked research on hereditary variation in these organisms, or at least made it more difficult.

In the first half of the twentieth century, the study of bacterial physiology had made great advances: bacteria were found to be able to adapt to new sources of food,[3] but it was not known whether this adaptation was genetic.

Between 1920 and 1940, there was a major unification of the life sciences. Several factors contributed to the acceptance of the idea that the same basic mechanisms explained the functioning of all forms of life. In the 1930s, the evolutionary synthesis succeeded in unifying Mendelian genetics, population genetics, and the Darwinian theory of evolution.[4] Models suggesting that bacteria contained no genes and were stable simply because the chemical reactions within them were in equilibrium had tended to separate micro-organisms and the rest of the living world.[5] This isolation of bacteria was all the more surprising given that viruses, which are lower on the scale of organic complexity than are bacteria, were considered to be almost pure genes.

Results from biochemistry played a key role in this unification of biology. In the 1930s, André and Marguerite Lwoff showed that all organisms use the same vitamins and coenzymes. The same metabolic pathways and the same transformations of biological molecules were found in bacteria and yeast, and in the muscles of birds and mammals. Beadle and Tatum's experiments had shown that each of the enzymes that catalyzed these essential metabolic steps was precisely controlled by a single gene; it was surely impossible that in bacteria these steps could be independent from genes.

At the same time, a growing number of physicists turned to biological research, hoping to discover the elementary principles of the functioning of life. Unaware of the complexity of the biological world, they were convinced that the same principles must be at work in a virus as in an elephant. For Delbrück, the mysterious replication of bacteriophages had to obey the same rules as the clearly genetic reproduction of higher organisms.

The first experiments carried out by the physicists to uncover the secrets of life reinforced their unifying vision. X-rays, which Hermann Muller had shown could alter genes, also perturbed bacterial development.[6] The same turned out to be true for ultraviolet rays and various chemicals. Similar doses and wavelengths acted in the same way on all forms of life.

A number of experiments had already suggested that bacteria showed hereditary variations. In 1934, I. M. Lewis proposed a novel experimental approach that would prove whether metabolic variations in bacteria were due to adaptation or to the selection of favorable mutants. All that was required was to measure the fraction of bacteria that could grow on a new metabolite *before* it was added to the medium, and to compare this value with that found *after* the metabolite had been added. If the two values were equal, metabolic variations must be hereditary. Lewis carried out two long experiments that confirmed his selective model.[7]

Luria and Delbrück's 1943 experiment thus took place in an extremely favorable context.[8] It also fitted in with their research on how bacteriophages replicated within bacteria. As noted, Luria and Delbrück had shown that "interference" occurred when two different phages were added simultaneously to the bacteria: one of the phages blocked the development of the other.[9]

To conduct this experiment, Luria had to determine the nature of the phages that managed to grow in the co-infected bacteria. This involved using bacteria that were resistant to one of the phages, obtained by pre-incubation with that phage. Most of these pre-incubated bacteria died, but some survived because they were resistant to the added phage. By picking out these colonies and allowing them to multiply, Luria found that he could grow bacterial cultures resistant to one or another phage. By adding phage of an unknown type to cultures with a known resistance profile, he could determine the nature of the added phage.

At this point, Luria was more or less obliged to investigate the origin of this bacterial resistance to bacteriophages. Did bacteria become resistant following their pre-incubation with the phage? Or were some of them already resistant prior to any contact with the phage, and these selected only later, during the infection period?

Luria's story is an interesting one. He was born in Turin in 1912 into a middle-class Jewish family.[10] Having begun the study of medicine without any particular enthusiasm, he was convinced by one of his friends that the new physics was the place to be, and he found his way to Enrico Fermi's laboratory in Rome. Luria quickly discovered that his mathematical training was not sufficient for conducting research in physics. But his stay in Fermi's laboratory was nevertheless a turning point in his life; there Franco Rasetti, who gave lectures on spectroscopy, told him about Delbrück's recent articles on the structure of the gene. Luria, who was fascinated by these articles, felt they addressed what he later called "the Holy Grail of biophysics." It was quite by chance that he heard about bacteriophages. One day he found himself in a stalled trolley bus seated next to the bacteriologist Geo Rita. They got to talking, and Rita invited Luria to visit the laboratory where he worked on bacteria and phage.[11]

Although he did not know it, Luria's conversion to phage took place at the same time as Delbrück's. In order to begin work, Luria had applied to the Italian government for a grant to go to Berkeley to study radiation biology. His grant was approved shortly before Mussolini proclaimed Italy's "racial" laws, but it was subsequently cancelled. Knowing what had happened in Germany on Kristallnacht, Luria quickly left Italy for Paris, where he was able to get a grant from the Fonds de la Recherche Nationale, the forerunner of the CNRS (National Centre for Scientific Research), to work on the effect of radiation on bacteriophage with the French physicist Fernand Holweck in the Institut du Radium, founded by Marie Curie. During his stay in Paris, Luria also worked with Eugène Wollman, a phage specialist at the Pasteur Institute.[12] Two days before German troops arrived in June 1940, Luria left Paris on a bicycle; he managed to get to Marseilles, where he obtained a visa for the United States.

With the help of the Rockefeller Foundation, and thanks to a scientific reputation based on his first articles, Luria was given a position at the College of Physicians and Surgeons at Columbia University. In late 1940, he met Delbrück at a conference of the American Physical Society at Philadelphia and the two men struck up a partnership that was to last many years, and which was renewed each summer at Cold Spring Harbor. In 1942, Luria worked with Delbrück at Nashville for a year, then in 1943 he obtained a permanent position at Indiana University at Bloomington. During his first two years in the United States, Luria helped Anderson take electron micrographs of the phage.[13] Shortly after his arrival in Bloomington, Luria carried out the famous fluctuation test experiment.

According to Luria's own account, the idea for this experiment came to him when he was looking at a slot machine during a dance at Indiana University.[14] When a coin is inserted into a slot machine, the gambler generally wins nothing. Every now and then, however, he or she hits the jackpot. Overall, no more coins are won than are inserted. But in terms of amounts paid out, the slot machine does not behave statistically: if it followed a

Poisson distribution, the player would recover one, two, or no coins each time, rarely more.

Luria applied this reasoning to his experiments on developing phage-resistant bacteria. Take twenty independent bacterial cultures, prepared from the same number of bacteria, and add phage. If phage do indeed induce resistance, the number of resistant bacteria will vary from culture to culture, but within reasonable "limits," according to a statistical Poisson law. If, by contrast, phage only help to select resistant bacteria, the number of such bacteria will vary considerably. It will be very large if the mutation that causes resistance appears during the first stages of bacterial replication. As controls, Luria used samples of the same culture. The result was several flasks of bacteria that, in the presence of phage, would give a variable number of resistant bacteria, distributed according to a Poisson law.

Luria wrote to Delbrück about the experiment and then began it without delay. Delbrück quickly replied, and then a week later sent him a mathematical analysis of the experiment,[15] showing that the results made it possible to calculate the *rate* of mutation, an essential measure for any genetic analysis. This was the first time that a mutation rate had been measured with such precision. The Luria-Delbrück experiment was a complete success and proved that the resistant strains were produced by mutation.

For many, this experiment marks the birth of bacterial genetics. In classical genetics, however, characters are defined and their transmission and recombination are followed over a number of generations. The Luria-Delbrück experiment had none of these characteristics. The experiment was important because it brought bacteria into the general understanding of evolution that had been outlined a few years earlier in the neo-Darwinian synthesis: organisms show variation, and variant organisms are selected over time.

The Luria-Delbrück experiment represented the triumph of Darwinism, a victory in the heart of what for many years had been "the last stronghold of Lamarckism"[16]—microbiology. This experiment linked Darwinism and molecular biology once and for all (see Chapter 21). Despite its success, however, the experiment did not immediately give geneticists access to the world of bacteria. Genes can be localized only when they recombine, reorganizing themselves into new groups, and the simplest phenomenon that leads to recombination is sexual reproduction.

At the time, it was accepted that bacteria reproduced only by fission. A few experiments had been carried out to detect sexual recombination, but they had all produced negative results, though a positive result would probably have been found only if there had been a very high recombination rate.

In *The Bacterial Cell,* published in 1945, René Dubos of the Rockefeller Institute showed that there was no strong experimental argument in favor of sexual reproduction in bacteria,[17] but that this absence of proof might simply show that the phenomenon had not been studied sufficiently. The unification of biology around the neo-Darwinian synthesis[18] and the important evolutionary role this view gave to sexual reproduction tended to suggest that this mode of reproduction must have originated in the simplest, most "primitive" organisms, such as bacteria.

Nevertheless, looking for the existence of sexual reproduction in bacteria was a risky business. The whole field of bacterial genetics was full of contradictory results and pitfalls. Any scientist who realized that the key to success is to attack "soluble" problems[19] would instinctively have avoided the subject.

The only scientists who can allow themselves the luxury of studying such a subject are those for whom the marginal costs of such an attempt are low, to borrow terminology from economics. This would include very young scientists, who can allow themselves a moment of rashness, or very famous scientists, whose reputation would not be harmed by failure. Joshua Lederberg and Edward L. Tatum formed just such a pair.[20]

Lederberg, born in 1925, was a medical student at Columbia University, and at the age of eighteen began work on *Neurospora* in Francis Ryan's laboratory. Ryan had just returned from a postdoctoral visit to Stanford, and had been inspired by Beadle and Tatum's experimental approach.

As soon as Avery's results were published, Lederberg tried to reproduce the transformation experiment in *Neurospora.* Although he failed, he had an idea for a new experiment to try to detect sexual recombination in bacteria. For this experiment, he required mutant bacteria, like the mutations obtained by Beadle and Tatum in *Neurospora,* which needed certain vitamins or amino acids to grow. Tatum had just begun work on bacteria and had obtained *E. coli* mutants that were just right for Lederberg's experiment.

Toward the end of 1945, after Tatum had left Stanford for Yale, Lederberg wrote to him explaining his project and asking for help. Tatum was busy moving and, with his change of subject, did not have much time to spare, but he strongly encouraged Lederberg and invited him to visit his new laboratory in March 1946.

In less than six weeks, Lederberg showed that his genetic markers were relatively stable and, after crossing, obtained recombinant bacteria, thus proving that genetic material had been exchanged and that some form of sexual reproduction must be taking place in bacteria. The result was presented for the first time at the Cold Spring Harbor international conference in July 1946,[21] where it was well received, with two notable exceptions: Delbrück, who thought the phenomenon uninteresting because it did not immediately lend itself to a kinetic analysis, and the French scientist André Lwoff, who suspected that the complementation observed between the various deficient strains of *E. coli* was not genetic, but was simply the result of an exchange of metabolites between bacteria. With Max Zelle's help, Lederberg isolated recombinant bacteria under the microscope and was able to answer Lwoff's criticism by proving that the mutations did indeed show genetic complementation. At the age of twenty-one, Lederberg had brought bacteria into the world of sexual reproduction and had made his own entry into the prestigious world of the young molecular biologists.

"What if" is generally a sterile question in history. But knowing what we know today, Lederberg's result appears to be a near miracle. He chose a strain (K12) of bacteria that was not widely used but that we now know was one of the few strains in which conjugation (mating) could be detected. If Lederberg had not been lucky, he would probably have had to test at least twenty different strains before getting a single positive result. After several negative results, would he have bothered to continue? Even after he selected K12, Lederberg's luck still held. By chance, he chose to follow the transmission of genetic markers developed by Tatum that all turned out to be on the same region of the *E. coli* chromosome. As Jacob and Wollman later showed, during conjugation chromosomal transmission begins at a precise physical point on the chromosome; the markers Lederberg used all happened to be close to this point, and were thus exchanged with a relatively high frequency.

In fact, genetic exchange in bacteria is a rare event of little physiological importance in normal bacterial life. For biologists, however, its discovery confirmed the fundamental unity of the living world. And most of all, it provided new tools for studying genes.

Over the next fifteen years, bacterial genetics made remarkable progress.[22] In 1945, phage genetics joined bacterial genetics, when Luria showed that just as there were mutant bacteria capable of resisting phages, so too were there mutant phages that could grow on these resistant bacteria.[23] At the same time, Hershey isolated other phage mutants that produced lysis plaques with different shapes.[24] In 1946, Hershey and Delbrück independently proved that if bacteria were simultaneously infected by two phages carrying different mutations, "recombinant" phages containing either both mutations or no mutations at all could be obtained.[25]

The power of bacterial genetics derives from the fact that it is possible to select extremely rare recombination events in relatively short spaces of time, through the use of antibiotics, phage, or—in the case of phage genetics—mutant bacteria. Seymour Benzer and a number of other researchers used these possibilities to take genetic analyses to a level well below that of the gene. Bacterial genetics thus helped bring the gene closer to the real world—and to molecules.

The contribution of bacterial genetics to the growth of molecular biology was particularly complex. Many important discoveries in molecular biology, such as the deciphering of the genetic code, took place outside the field of bacterial genetics. And yet the tools developed by bacterial geneticists often made it possible to find an elegant verification or demonstration of results that had been obtained through biochemical studies.

Bacterial genetics made its biggest contribution to the growth of molecular biology in the field of gene regulation, in particular in the work of the French school of molecular biology (see Chapter 14). From a longer-term perspective, bacterial genetics also provided the essential tools for the development of genetic engineering (see Chapter 16).

Bacterial genetics is at least as "abstract" and formal as *Drosophila* genetics, but it is also thoroughly practical, based as it is on a set of simple and precise techniques for manipulating bacteria and phage. Although bacterial genetics is characterized by the fact that gene transfer is always partial, the powerful selection techniques developed by Lederberg and others largely overcame this problem.

As soon as bacterial recombination had been discovered, Lederberg tried to establish a genetic map of *E. coli.* Unfortunately, the data could not be easily interpreted. In 1951, Lederberg and his colleagues finally proposed that the bacterial chromosome was a branched structure with four arms.[26]

The Irish scientist William Hayes discovered that sexual reproduction in bacteria was extremely unusual, and his results provided scientists with a new interpretative framework. In 1952, Hayes showed that the transfer of different genetic markers during bacterial conjugation was neither simultaneous nor symmetric: one bacterium behaved like a donor (male), while the other acted like a recipient (female).[27]

The real nature of bacterial sexual reproduction was finally revealed by Elie Wollman and François Jacob. Wollman was the son of Eugène and Elisabeth Wollman, who during the interwar years had worked on lysogeny at the Pasteur Institute in Paris (see Chapter 14); he had been a student of Lwoff and from 1948 to 1950 had worked in Delbrück's laboratory at Caltech on phage T4 adsorption.[28] To study conjugation, Wollman and Jacob used a bacterial strain called Hfr ("high frequency of recombination"), which showed high levels of conjugation. Mating was interrupted by violently agitating the bacterial culture in a kitchen blender—the same apparatus (and principle) used by Hershey and Chase, and by Anderson before them, to separate bacteria from infecting phage. Wollman and Jacob showed that gene transfer took place regularly over time.[29] They hypothesized that the order in which genes were transferred was related to their position on the bacterial chromosome and that this order could be directly converted into a genetic map.

Following this discovery, bacterial conjugation became a powerful genetic tool. Contrary to the initial hypotheses, it turned out to be very different from sexual reproduction in other organisms. But at least in terms of the use that geneticists could make of it, it was similar to generalized transduction, discovered by Norton Zinder and Joshua Lederberg in 1950.[30]

The initial objective of Zinder and Lederberg was to show the existence of conjugation in bacteria other than *E. coli.*[31] The bacterium studied was *Salmonella.* Studies with antibodies had revealed that there were several antigens on the surface of this micro-organism, and that different strains of bacteria carried different combinations of these antigens. Lederberg felt that this complex antigen profile was the result of sexual recombination between strains.

Initial experiments suggested that there was indeed a transfer of genetic material between different strains of *Salmonella,* but that this transfer did not require the two strains to be in physical contact, as in conjugation. Genetic transfer was carried out by an unknown agent that could pass through filters and was not affected by the action of DNase. This new phenomenon of genetic transfer was thus different from Avery's transformation (see Chapter 3). The properties of this agent turned out to be identical to those of the bacteriophage,[32] which passively transported a fragment of genetic material from one bacterium to another.

When similar effects were discovered in *E. coli,* transduction became a very effective tool for studying the structure of genes and their chromosomal organization. Beadle and Tatum had shown that each step in the metabolic pathways of biosynthesis or degradation was catalyzed by a different enzyme, corresponding to a separate gene (See Chapter 2). Transduction experiments revealed that in bacteria, the genes involved in a given metabolic pathway were found on the chromosome in the same order as they intervened in the pathway.[33] This spatial organization means that genes can be coregulated.

Transduction nevertheless remained mysterious. How did bacterial genetic material become associated with the phage? The answer came indirectly, from the study of lysogenic phages. It was known that such phages could be integrated into the bacterial genome at specific positions; they could also transduce the host's genetic material, but the only genes that became associated with the phage were those situated near the site where the phage had integrated itself into the bacterial chromosome. Transduction thus corresponded to an imprecise "excision" of the phage, which took part of the surrounding genes with it when it left the chromosome. The simplest hypothesis that could explain generalized transduction was that transducing phages could also integrate themselves into the chromosome,[34] but with no spatial "preferences." Upon excision, they would take with them bacterial genetic material from wherever on the chromosome they happened to be.

Conjugation and transduction made it possible to draw up a genetic map of *E. coli* that had a higher resolution than those of classic organisms such as *Drosophila.* These new techniques thus helped genetics "descend" to the molecular level.

CHAPTER 6

The Crystallization of the Tobacco Mosaic Virus

The virus was the most appropriate organism for studying the chemical nature of the hereditary material. Viruses had been discovered at the end of the nineteenth century and characterized at the beginning of the twentieth century. Because of their minute size, they appeared to lie between the world of chemistry—molecules and macromolecules—and the world of biology, the simplest representatives of which were bacteria.

Most important, viruses had the same capacity as organisms and genes: they could reproduce and switch to a new stable state, that is, mutate. The fact that viruses might be genes, or primitive forms of genes, had been suggested by Hermann Muller in the 1920s, and was widely accepted by geneticists and virologists.[1]

Viruses appeared to be an ideal material both for studying genes and for tackling the problem of the secret of life. In addition to these scientific reasons, a number of practical factors made the study of viruses a key research topic: viruses are responsible for serious pathologies in humans, and also for diseases in agricultural animals and plants. Large sums of money have always been available for research on viruses. President Franklin Delano Roosevelt, who contracted poliomyelitis at age thirty-nine, was instrumental in securing funds for research on the polio virus and the development of a vaccine.

Viruses will not reproduce in synthetic media. This problem was overcome only by the development of the cell culture technique in the 1950s, which explains why, for many years, the structural study of animal viruses was extremely difficult. Although the study of bacteriophages was much easier, the precise experimental conditions required for studying their reproduction were developed only in the 1940s.

Throughout this period, plant viruses were the model system for studying viral reproduction.[2] The leaves of the infected plant were the equivalent of cell culture dishes: the viruses could be placed on them, and their multiplication directly measured by the growth of dead areas on the leaves.

The purification and crystallization of the tobacco mosaic virus (TMV), carried out by Wendell Meredith Stanley in the summer of 1935,[3] constituted a revolutionary event, widely reported in the media of the time.[4] Stanley was compared to Pasteur, and in 1946 he was awarded the Nobel Prize for chemistry.

This success was representative of a physico-chemical approach to the study of biological phenomena that was particularly well rooted at the Rockefeller Institute. Stanley was trained as a chemist, and had done his doctoral thesis on the synthesis of bio-organic compounds. In 1931, after a year-long stay in Munich, he was recruited by the director of the Rockefeller Institute, Simon Flexner, to work in the new laboratory of the Department of Animal and Vegetable Pathology, located in the institute's annex at Princeton.

Stanley found the atmosphere at Princeton extremely favorable for a number of reasons,[5] one of which was financial: the Rockefeller Institute was one of the research bodies to weather the budget restrictions that followed the crash of 1929. The atmosphere was also favorable scientifically: one of Stanley's colleagues at Princeton was John Northrop, who, after training in Morgan's group, had decided to turn to physical and chemical biology rather than genetics, because he felt that these disciplines were scientifically "safer." A year before Stanley arrived at Princeton, Northrop succeeded in crystallizing an enzymatic protein, pepsin.

But above all, Stanley found that the atmosphere at the Rockefeller Institute and at Princeton was "philosophically" favorable. As we have seen with regard to Avery, the dominant biological philosophy at the Rockefeller Institute was reductionist. Researchers sought to explain biological phenomena in physico-chemical terms. Sumner's success in crystallizing the first enzyme, urease, and Northrop's recent results had both strengthened confidence in this reductionist approach, first championed in the 1920s[6] by the German biologist Jacques Loeb, who was also Northrop's mentor.

Stanley was not the first person to try to purify and crystallize the TMV, but he brought to the project his substantial skills as a chemist and, most important, his conviction that the virus was a protein and that methods for purifying proteins would enable him to isolate the virus in a pure form. Having clearly shown that the infectivity of the TMV was destroyed by pepsin, by extreme pH levels, and by a number of chemicals known to inactivate proteins, he required only a few weeks to complete the purification and crystallization of the virus. In his article published in *Science*,[7] Stanley stated that the crystalline material he obtained was a protein, and that its properties remained unchanged after more than ten successive crystallizations.

In the years that followed, Stanley discovered more about the properties of the virus. He gave a sample to the Swedish chemist Theodor Svedberg while Svedberg was on a visit to New York. On his return to Uppsala, Svedberg used ultracentrifugation to show that the molecular mass of the virus was seventeen million daltons and that it was rod-shaped. This structure was confirmed in 1940 in one of the first studies carried out with the electron microscope built in the RCA laboratories at Princeton, close to Stanley's lab. Stanley's laboratory was also one of the first in the United States to be equipped with a Tiselius electrophoresis machine (see Chapter 9). Stanley used the new machine to characterize different strains of the TMV as well as other plant viruses.

Stanley's finding was received enthusiastically. Without a doubt, it was the most important discovery to have been made at the already prestigious Rockefeller Institute. The press gave Stanley's work a great deal of coverage, to which he himself made no small contribution. But it was the crystallization of the virus, rather than its purification, that most excited the imagination. A crystal was a symbol of the material, mineral, mechanical world. As the *New York Times* put it: "in the light of Dr. Stanley's discovery, the old distinction between life and death loses some of its validity." In a way, the crystallization of the TMV represented life's "twilight zone."[8]

It was not long before Stanley's work was criticized. Some of these criticisms dealt with matters of principle—crystallization does not mean purity—others with the experiments themselves. In Britain, Frederick Bawden and Norman Pirie tried to reproduce Stanley's results. They managed to crystallize the virus, but found that 6 percent of it was composed of a nucleic acid (RNA).[9] Furthermore, they showed that any treatment that inactivated this RNA also changed the infectivity of the virus. They concluded that the virus was not a pure protein but a nucleoprotein. This result fitted in with those models that considered nucleoproteins to be the essential constituent of genes and chromosomes. Stanley never replied directly to his critics. He later admitted that RNA was indeed a constituent of the TMV, but did not accord this observation any great weight.[10] It is true that, in 1940, the structure of RNA was as little known as that of DNA, and this did not encourage researchers to think that it could play a role in hereditary transmission.

Thus the purification and crystallization of the TMV did not advance the characterization of the chemical nature of the genetic material. In 1956, when there was no longer any doubt that nucleic acids carried hereditary information, Gerhard Schramm and Heinz Fraenkel-Conrat finally provided decisive proof that the RNA in the TMV was indeed the infectious material. Avery's experiment thus remained the first to show convincingly that the genetic material was formed of a nucleic acid, despite the fact that the link between transformation and gene expression was a lot more tenuous than the link that was thought to exist between viruses and genes.

To leave the matter there, however, would be to underestimate the importance of Stanley's experiment. On a philosophical level, it helped to validate the reductionist, physico-chemical approach to biological phenomena. Scientifically speaking, it had an enormous impact, in particular on geneticists. Muller welcomed the result enthusiastically. Northrop reoriented part of his work to isolating the bacteriophage. Stanley himself devoted more and more time to genetics and encouraged some of the researchers in his laboratory to enter this new field. Bit by bit, the gap separating geneticists and biochemists began to close. The techniques Stanley used to purify and above all to characterize "his" virus became the experimental reference for virologists.[11] The Rockefeller Institute was equipped with a new ultracentrifuge in order to prepare large quantities of virus. This new centrifuge not only permitted the preparation and purification of viruses on a large scale, but, more important, opened the way to the fractionation and isolation of cellular constituents. These studies rapidly led to the discovery of microsomes, intracellular particles upon which protein synthesis took place. But it would be another twenty years before the relation between genes and microsomes was finally clarified (see Chapter 13).

CHAPTER 7

The Role of the Physicists

As we have seen, a number of physicists played an important role in the birth of molecular biology despite having no biological training. This might not seem surprising, given that many of them used the technology of physics to study biology. But strangely enough, these physicists did not work very closely with the biologists; instead, they chose their own research projects and developed the biological "spin-offs" of their results.

The movement of physicists into biology involved far more than a handful of "specialists."[1] Many scientists with long and distinguished careers in physics decided to retrain completely and begin studying biology: one such example was the Hungarian physicist Leo Szilard.[2] Szilard was the first person to provide a "physical" explanation of Maxwell's demon and thus founded what was to become information theory. As the inventor of the atomic chain reaction, he played a key role in setting up the Manhattan Project to develop the first atomic bomb: with two other Hungarian physicists, he alerted Einstein to the risk that Nazi Germany might develop an atom bomb, which led Einstein to write to President Roosevelt. But at the beginning of 1945, Szilard was also one of the first scientists to oppose the military use of the bomb. At the end of the war he turned to the study of bacterial metabolism and played an important role in the analysis of gene regulation in bacteria (see Chapter 14).

George Gamow, one of the fathers of the "big bang" theory of the origin of the universe, was another physicist who made important contributions to molecular biology. Gamow reacted quickly to Watson and Crick's article on the double helix structure of DNA and, in the days that followed its publication, wrote to them proposing a code that would explain the relation between the sequence of bases in DNA and amino acid chains (see Chapter 12).[3] Gamow's model and its errors will be discussed later; for the moment it is sufficient to note that he was the first to try to decipher the genetic code using a theoretical approach. Gamow's letter came as a shock to Watson and Crick—despite the fact that they had used the term "code" in their article, they had not taken their model to its logical conclusion and imagined that it would be possible to decipher the code. Gamow later made many other contributions to the theoretical study of the structure of the genetic code.[4]

Many other physicists were fascinated by biology. Can their role be "quantified?" Nicholas Mullins has carried out a detailed study of the phage group,[5] providing data on the scientific origin of the various researchers involved in the group between 1945 and 1966, as measured by the discipline in which they had obtained their doctorates. Prior to 1945, three of the six members of the group had obtained a Ph.D in physics or chemistry. Between 1946 and 1953, ten of the nineteen new members had a Ph.D in physics, biophysics, or chemistry. The "weight" of physicists (and chemists) began to decrease only in 1954 (four out of thirteen new members between 1954 and 1962). This reduction was no doubt due to the impact of the discovery of the DNA double helix (see Chapter 11), which helped to attract young biologists to molecular biology at a time when it was becoming a separate science with a large number of potential research subjects for graduate students.

This attempt at quantification is based only on the phage group, which was only one of the many branches of molecular biology. Furthermore, the structure of the phage group was sufficiently vague for its exact composition to be difficult to determine; data relating to it should thus be treated with caution.

Whatever its precise value, the numerical importance of the physicists cannot be denied. François Jacob has provided a particularly penetrating analysis of the reasons for this intellectual migration and the changes that took place as a result of the Second World War.[6] Prior to the war, some young physicists were disappointed by the discipline in which they were training. It was not so much that physics, basking in the glow of the double revolution of quantum mechanics and relativity, was not a brilliant subject: each day these new theories led to new discoveries. Rather, physics had entered a phase of what Thomas Kuhn[7] called "normal science," in which most activity consists of "puzzle-solving" rather than questioning the very bases of the discipline. For the most ambitious young physicists, checking or at best slightly improving their elders' models was somewhat uninspiring. Furthermore, the very structure of research in physics was changing. Isolated researchers were being replaced by teams focusing on big projects or massive pieces of apparatus such as particle accelerators. In this kind of work, based on the collaboration of a number of specialists, the role and contribution of each individual are reduced, or at least more difficult to determine. Some scientists are too individualistic to want to be virtually anonymous members of a large multidisciplinary team.

Beginning with the Second World War, other motives tended to push some scientists away from physics. Many American and British physicists were closely involved in the war effort. For the first time in history—with the exception, perhaps, of the role of scientists in the organization of the armies of the First Republic in France—physicists and mathematicians were called upon to help the military. They developed radar and sonar, improved communications, decoded messages, and developed powerful calculators (of which computers were later a natural development[8]). It would be tedious to list all the major technological advances that were made during the Second World War by brilliant physicists working for the defense of their countries. Above all, the Second World War was a war of intelligence, of organization, and of science. The British were the first to realize the new form that the war had taken.[9] A key reason for this was no doubt the fact that their geographical and military situation left them no alternative; their only hope of victory was to shift the terrain of battle.

The teaming up of different scientists and the importance given to information and its exchange led to a new vision of the world. But a negative consequence of scientists' involvement in the war effort was the desire felt by many of them to escape from what they saw as a form of conscription. They wanted to put an end to the guilt they felt at being associated with various military operations at the end of the war, which fundamentally altered what, for a long time, had seemed to be a struggle between good and evil. For example, the allied bombardment of Dresden, which was presented as a response to the bombardment of Coventry, produced many more innocent victims. Unlike the Manhattan Project, the atomic bombs dropped on Hiroshima and Nagasaki could not be justified by a potential German nuclear threat, and reminded physicists that the military and politicians had motives and objectives that were different from theirs and which they could not influence. Research on the atom and on fundamental physics came out of the affair sullied and, for many years, suspect. Biology appeared as a new field, far from political concerns and sheltered from potential military uses.

Physicists were also attracted to biology because it seemed to harbor a large number of unsolved fundamental problems and to be the "new frontier" of scientific knowledge. Quantum physics, and the new chemistry it had produced, appeared to be able to provide the tools and concepts necessary for understanding the mysteries of biology. In fact, it was the fathers of quantum mechanics, Niels Bohr and Erwin Schrödinger, who were the heralds of this new approach. An analysis of their writings will make clearer the motivations and hopes they shared with many physicists.

Of all the authors of the quantum revolution, Niels Bohr is probably the best known to the general public. All high school students learn about Bohr's conception of the atom, with its electrons orbiting on precise trajectories around a positively charged nucleus. His description of the position of the electrons around the nucleus—which went against the dogma of classical physics—was able to explain experimental data on the absorption and emission of light by atoms and laid the basis for quantum mechanics.

Bohr did not, however, play a direct role in the birth of quantum mechanics. He was not involved in the wave approach developed by Louis de Broglie and Erwin Schrödinger, or in the more mathematical studies of Werner Heisenberg, Paul Dirac, and Wolfgang Pauli. And he did not contribute to Schrödinger's synthesis. But their results justified his atomic model *a posteriori.* Indeed, Bohr was the first to see the radical novelty of the new theories, which put into question the absolute determinism of classical physics and justified different, complementary approaches to reality, abolishing the division between the observer and the observed world. A small group was formed around Bohr at Copenhagen, pushing the consequences of the new theory to their limits. Many physicists rejected what became known as "the Copenhagen interpretation." Neither Einstein nor Schrödinger would accept the abandonment of determinism; for the rest of his life, Einstein searched for the hidden variables that would explain the apparent indeterminisms of the new physics.

Bohr was also interested in biology. He was the son of Christian Bohr, a famous Danish physiologist who had studied the fixation of oxygen by blood. As we have seen, Bohr encouraged many physicists, such as Delbrück, to turn to biology. Bohr invited George Hevesy to the Institute of Theoretical Physics, which he had founded in Copenhagen, and encouraged Hevesy's research into the use of radioactive isotopes as biological markers,[10] including asking for support and funding from the Rockefeller Foundation. After the war, this new technology was to be of major importance for the development of biochemistry and molecular biology.

In August 1932, Bohr was invited to give the inaugural lecture at the International Congress on Light Therapy at Copenhagen. The following year, this lecture was published in *Nature.*[11] The title of Bohr's talk—"Light and Life"—gives an idea of its content and importance. After a long presentation of quantum mechanics, Bohr turned to the essential theme of his lecture: how these recent results on the nature of matter (and of light) change our vision of life. Bohr's objective was not to make a scientific transfer from physics to biology by applying results from one science to the other, but to make an "epistemological transfer," to try to see how the new vision of the physical world changed perceptions of the biological world.

One of the key results of the new physics was the principle of complementarity. A given object, such as a photon, could, indeed should, be studied by approaches that were different but complementary—it had to be studied both as a wave and as a particle. In the same way, Bohr argued, organisms should be studied by complementary approaches. No one could deny the importance of chemical and atomic phenomena in the functioning of all organisms. The strides made in the study of biological molecules had quashed the idea that there was a "vital force," and suggested that living matter was no different from inorganic matter. Nevertheless, studies of organic molecules required that the organism that contained them be killed—the reductionist study of life meant that life had to be destroyed. To resolve this paradox, Bohr proposed to accept the phenomenon of life as an elementary fact that could not be explained, the biological equivalent of a quantum of action, something that had to be admitted in order to explain everything else. Bohr asked if it was possible to imagine and even develop another biology alongside the reductionist approach that tended to destroy life, one that would consider life an irreducible phenomenon and reserve an essential place for its teleological, intentional aspects.

In his conclusion, Bohr tried to correct the perhaps negative impression given by his lecture. The limits of knowledge of the inanimate world had not prevented quantum mechanics from leading to a far superior understanding of nature and to a massive increase in humankind's power over it. Similarly, renouncing any attempt at explaining "life" would not be an obstacle to making enormous progress in the future understanding of biology.

This lecture has often been cited but rarely read, and it seems that those who have read it have misunderstood it. Bohr's ideas, and even more so those of Delbrück, who developed Bohr's positions and used them as the basis of his biological research program, were dismissed by their critics as wrong. Bohr had suggested that physics would never be able to explain the functioning of organisms, and that other principles would have to be discovered before the biological world would become comprehensible. Today, the reductionist approach seems increasingly complete—conversely, the need for other approaches, the search for new principles, seems less and less justified. As we have seen, this search was one of Delbrück's motivations, which explains why he turned away from molecular biology at the point where Watson and Crick reduced the self-replicative abilities of the gene to the complementary chemical structure of the double helix. Did the same hope of discovering new principles also motivate Bohr? In fact, he was more than somewhat reserved when Delbrück tried to present him as the inspiration for his own research.[12]

This apparent misunderstanding between Delbrück and Bohr suggests a different reading of Bohr's Copenhagen lecture. Bohr was no stranger to the idea that life could not be reduced to the material world, nor mind to matter. But surely the suggestion of a complementary approach to biology implied a strategic prudence rather than an antireductionist manifesto. Because it rejected all philosophical reductionism, a declaration that the irreducibility of life is the quantum of the new biology opened the road to experimental reductionism. Leaving aside all polemical interpretations, by limiting the philosophical ambitions of the chemical, physical, or biological approaches, scientists can proceed "as if" life were precisely nothing other than the result of interactions between the molecules of the organism.

In the final analysis, Bohr's motives do not matter. What does matter is that his lecture helped to turn the attention of physicists toward biological "objects."

In a series of lectures given in Dublin in 1943, and in his book *What Is Life?*,[13] which was based on these lectures and was published in 1944, Schrödinger drew the attention of young physicists to the results of genetics and suggested that quantum mechanics would be able to explain them. No sooner was Schrödinger's book published than it was taken—unlike Bohr's lecture—to represent a transfer of the new concepts of physics into biology, a takeover bid by physicists on the mechanisms of heredity.

The plan of the book shows Schrödinger's intentions. He begins with a description of the principles upon which order is based in classical physics. For the physicist, thermodynamics—order at the macroscopic level—is merely the statistical result of microscopic disorder. Schrödinger then goes on to give a description of the main results of genetics. He underlines the stability of genes and the fact that it is perturbed only by sudden and rare mutations, which then lead to new stable states. Schrödinger describes Muller's success in inducing mutations by radiation. These two properties of genes—stability and mutability—cannot be explained by classical physics. Genes are far too small for their stability to be the result of a statistical equilibrium of their constituent molecules. Their properties evoke the stable energy levels that quantum mechanics showed exist in molecules. The molecules that form genes would simply have to be sufficiently complex for the energy levels that they could attain to have the observed degree of stability. Schrödinger then shows how Delbrück, in using this molecular model of the gene, had been able to explain some of the characteristics of mutations and had tried to measure the size of genes. The book continues with two chapters on order in organisms, and closes with an epilogue on determinism and the problem of free will.

Schrödinger's book was a remarkable success. Many of the founders of molecular biology claimed that it played an important role in their decision to turn to biology. Gunther Stent, a geneticist (and a historian of genetics), has argued that for the new biologists it played a role like that of *Uncle Tom's Cabin.*[14] Schrödinger presented the new results of genetics in a lively, compelling way—much better than the biologists had. Fifty years later, the book has lost none of its seductiveness: its clarity and simplicity make it a pleasure to read.

A modern molecular biologist would feel quite at home reading Schrödinger's book. He or she would share Schrödinger's determinist vision of the role of genes: "In calling the structure of the chromosome fibers a code-script we mean that the all-penetrating mind, once conceived by Laplace, to which every causal connection lay immediately open, could tell from their structure whether the egg would develop, under suitable conditions, into a black cock or into a speckled hen, into a fly or a maize plant, a rhododendron, a beetle, a mouse or a woman."[15]

Despite the problems raised by the inclusion of "woman" in the list, today's molecular biologist would wholeheartedly agree with this research program. Furthermore, Schrödinger was one of the first to use the word "code" to describe the role of genes. In another part of the book, where he attempts to describe the structure of genes, Schrödinger puts forward the hypothesis that they are formed of some kind of aperiodic crystal—a remarkable anticipation of the nonmonotonic polymer structure of chains of nucleic acids.

The idea that Schrödinger's view was original has been vigorously contested,[16] and most contemporary historians and philosophers of science are ill at ease with the idea of "precursors" or "founding fathers." Philosophers of science—often rightly—consider this kind of position a consequence of scientists' developing their own historical mythology. Schrödinger's book was in fact fully appreciated only in the 1960s, when the key findings of molecular biology had already been discovered. Perhaps this is a case of the new science's constructing its own historiography, furnishing itself with such prestigious founders as Niels Bohr and Erwin Schrödinger.[17]

It can be shown that Schrödinger did indeed borrow a number of ideas from other scientists—from Delbrück, but also from certain geneticists such as Muller. It is also easy to point out Schrödinger's blind points: he was not interested in the chemical nature of genes, and he did not mention Avery's work on the transforming factor. And though he cited Delbrück, it was for his work on the gene, and not as the leader of the phage group, which at the time was growing rapidly. Finally, Schrödinger did not discuss the exact role of genes; nor did he mention Beadle and Tatum's work on genetic physiology.

None of these criticisms, however, takes away from Schrödinger's originality, which flowed from the physicist's farsighted vision, allowing him to see genes merely as containers of information, as a code that determines the formation of the individual.[18] Schrödinger dared to say what no geneticist would have said, that "these chromosomes . . . contain in some kind of code-script the entire pattern of the individual's future development and of its functioning in the mature state."[19] For Schrödinger, genes were no longer merely guarantors of order within the organism, mysterious conductors that ensured its harmonious functioning. Instead, they were scores that, down to the smallest detail, determined the functioning and the future of every organism. To decode the information contained in chromosomes is to know the organism. Schrödinger pushed to its logical conclusion the geneticists' view that genes were the heart and soul of the organism.[20] He thus anticipated the results of molecular biology that were to show how genes determine the position and nature of all amino acids, and thus of all proteins within the cell.

Schrödinger's vision was extremely original. Nevertheless, this originality was not necessarily perceived by contemporary readers—a survey of the various reviews of Schrödinger's book in journals of the time is particularly revealing in this respect.[21] Most of the reviewers concentrated on the final chapters of the book, devoting most of their attention to his thoughts on order, on entropy, on what Schrödinger called neguentropy—the property of living beings that enables them to increase order at the cost of their surrounding environment—and on the problem of free will. But the reviews do not even mention the idea of a code, of genes being constituted by an aperiodic crystal, or the new conception of the role of genes. Furthermore, when Schrödinger's readers were interviewed twenty years after the book appeared, none of them remembered having been struck by the novelty of the ideas it contained, but instead claimed to have been seduced by its clarity, by the fact that it made biology attractive to young physics students.

The failure of early reviewers to recognize Schrödinger's real influence is not, however, the end of the story. It simply shows that Schrödinger's book was not, strictly speaking, a scientific work. It was neither a treatise on biology nor a collection of articles. Its aim was not to encourage new experiment nor to present new models of the functioning of organisms, but rather to offer a new vision of biology. But a new vision does not impose itself as a new theory can. At first it goes unnoticed, and it is never the work of one person. In this respect, Schrödinger was as much the representative of this vision as its author.

To say that this new vision had no direct influence on scientists or on the experiments they carried out does not mean that it had no influence at all. Schrödinger and the new concepts of information, of code, and of program, were not the originators of the new science of molecular biology. But they did provide a conceptual framework within which experiments were worked out and interpreted, and which provoked new experiments and new research projects.

Although many other physicists were also interested in biology, few of them went so far as Schrödinger—giving lectures aimed at the lay reader and publishing them as a book. The reasons for Schrödinger's dedication can be found in his personality, his academic training, and his philosophical ideas.[22] Schrödinger was one of the founding fathers of quantum mechanics, but he refused to accept Bohr's nondeterminist explanation. His final work was devoted to an attempt to unify the separate branches of physics. Restoring determinism and reuniting the different domains of physics seem to have been his two key aims throughout his career. In this respect, he was faithful to Ludwig Boltzmann, who, although he was not Schrödinger's teacher, had been his model at the University of Vienna between 1906 and 1910. Boltzmann had unified two branches of physics—thermodynamics and mechanics—that had been separated throughout the nineteenth century, and had shown that there is physical order behind the apparent disorder of atoms.

Although Boltzmann's influence can help us understand Schrödinger's desire to find a principle of order at the center of life, Schrödinger's personality and the environment in which he grew up also explain his highly original approach to biological problems. In 1925, before he made his key discoveries in physics, Schrödinger wrote a philosophical book in which he exposed his "conception of the world."[23] This book reveals the philosophical framework within which his scientific thinking was situated.[24]

An adept of Hindu philosophy, Schrödinger refused to accept the dualism of life and the world. Each living being, he believed, is merely a facet of the same totality: this explains both how we can all know the world and that this knowledge is the same for each of us. He claimed that organisms differ from inanimate matter only by the existence of memory: the memory of past events, and also the memory of previous generations. Instincts are the mnemonic trace of the behavior of the organisms that preceded us. Embryonic development is also the memory of the evolution of life. Although he did not use the term, for Schrödinger the existence of a genetic memory constituted the specificity of all life.

This early work thus contains the hazy outline of the research program Schrödinger was to propose to his physicist colleagues some twenty years later: find a principle of order in organisms and link it to the memory of things past. These two themes—memory and the search for a deep, hidden order—were at the heart of early-twentieth-century Viennese ideas.[25] Thanks to Schrödinger, molecular biology can be seen as a late flowering of the intellectual effervescence that characterized Vienna in the early years of the century.

Today, very few physicists or mathematicians dare to study biological problems, in contrast with the situation in the first half of the twentieth century. The growing specialization of the sciences is not the only factor responsible for this change. Early in the century, it was commonplace to think that biological knowledge lagged behind that of the inanimate world. Bridging that gap, and using the recently acquired knowledge in physics and chemistry to do so, was a "natural" project, at least for the physicists. Biology appeared to be the new frontier of knowledge.

But the history of science is not only the history of ideas and desires; it is also the history of techniques and financial means. The desire to regenerate biology through the tools of physics and chemistry also guided the Rockefeller Foundation in its funding of research.

CHAPTER 8

The Influence of the Rockefeller Foundation

One of the most widely discussed questions in the history of molecular biology is whether research grants provided by the Rockefeller Foundation played an essential role in the birth of that science.[1] The bitterness of the debates over this question cannot be explained simply by the desire of the protagonists to understand the development of molecular biology. The stakes are much higher: there are very important implications for science policy. If the Rockefeller Foundation did play an essential role, this shows that scientific research can be oriented in a given direction from the "outside." If, by contrast, molecular biology underwent a more or less autonomous development and was simply supported by the Rockefeller Foundation, that implies that the development of scientific research is largely independent of external influences, and that science policy should aim to provide nonselective aid to research.

In contrast with many European countries, in the United States foundations have played an important role in the development of science.[2] The Rockefeller Foundation was created in 1913 by John Rockefeller, the famous petroleum magnate, to contribute to the well-being of humanity. This aim, it was held, could and should be attained by a systematic application of scientific knowledge.

Within this rather general framework, the aims of the foundation were nevertheless relatively diverse. From 1913 to 1923, the foundation was devoted to general education and public health. From 1923 to 1930, it mainly supported medical and scientific teaching. After the Second World War an important part of the Rockefeller Foundation's work was devoted to the development of high-yield strains of wheat and their distribution in Mexico, India, and other third-world countries. The period that concerns us here stretched from the 1930s to the years that followed the Second World War, during which the foundation underwent a major change of orientation.

From 1923 on, the grants given by the foundation for medical and scientific teaching increased at a slower rate than public funding. The foundation's contribution became less and less important, thus undermining its role as a major force. A new policy was necessary. This policy was the result of intense discussions that took place after the onset of the Great Depression of 1929. This economic crisis, one of the most serious that America had ever known, came after a long, uninterrupted period of growth. The scale of the slump implied a profound, noncontingent cause. For the Rockefeller Foundation administrators, the origin of the crisis lay in the gulf between humankind's understanding of the productive forces, which had grown continuously in the previous period, and its understanding of itself, which had not progressed. A quote from Warren Weaver, a mathematician and physicist by training, the head of the natural science division of the Rockefeller Foundation from 1931, illustrates the administrators' analysis of the crisis and the objectives of their new policy. "Our understanding and control of inanimate forces has outrun our understanding and control of animate forces. This, in turn, points to the desirability of an increased emphasis, within science, on biology and psychology, and on the special developments in mathematics, physics, and chemistry which are . . . fundamental to biology and psychology."[3]

Weaver and the other administrators gradually drew up a program that was initially extremely ambitious. Until 1934 an important aspect of Weaver's program was the understanding of humankind through the study of psychology, hormones, and nutrition. Weaver wrote:

> Can man gain an intelligent control of his own power? Can we develop so sound and extensive a genetics that we can hope to breed, in the future, superior men? Can we obtain enough knowledge of physiology and psychobiology of sex so that man can bring this pervasive, highly important, and dangerous aspect of life under rational control? Can we unravel the tangled problem of the endocrine glands, and develop, before it is too late, a therapy for the whole hideous range of mental and physical disorders which result from glandular disturbances? Can we solve the mysteries of the various vitamins . . . ? Can we release psychology from its present confusion and ineffectiveness and shape it into a tool which every man can use every day? Can man acquire enough knowledge of his own vital processes so that we can hope to rationalize human behavior? Can we, in short, create a new science of Man?[4]

In 1934, after the program had been examined and its psychobiological aspect criticized, Weaver refocused his effort on fundamental biology and the application of the new techniques of physics to biochemistry, cellular physiology, and genetics. The importance given to endocrinology and to the study of vitamins decreased, then completely disappeared from Weaver's jurisdiction.

Weaver was the first to use the term "molecular biology," in 1938.[5] This term described the new approach to biological problems using the techniques of physics and chemistry. Warren Weaver is thus the father of molecular biology, even if his use of the term did not correspond to today's definition.

For the period 1932–1959, the foundation contributed an estimated $25 million to molecular biology. Part of this money was given in the form of grants to young or experienced researchers to pay, for example, for visits by European scientists to U.S. laboratories, and part was given to fund specific research projects. This controlled attribution of grants broke with the previous policy of the foundation: in the past, important sums had been given to prestigious bodies that would themselves distribute the money and oversee its use. The distribution of limited sums for specific projects tended to encourage the acquisition of new pieces of equipment rather than the development of new research groups.

The money provided by the foundation played an important role in equipping many laboratories with spectrophotometers, ultracentrifuges, material required for X-ray diffraction, Geiger counters to detect radioactive isotopes, and so on. The Rockefeller Foundation also contributed to the development of these new technologies (see Chapter 9). Finally, the foundation continued to give substantial sums of money to long-term projects proposed by a handful of high-ranking laboratories, such as Linus Pauling's laboratory at Caltech and William Astbury's laboratory in Leeds, England.

The way in which these grants were awarded was also new: the funding decision was made by Weaver and a handful of his close colleagues. The quality of the project and of the principal investigator was measured by a network of correspondents, chosen from the most eminent scientists in the country concerned. This informal network provided Weaver with a massive amount of information, and was particularly helpful when it came to estimating how the money had been used and the scientific quality of the research that had been carried out thanks to the foundation's funding.

European laboratories benefited from this new grant-awarding policy. On the one hand, the foundation sought to restore the balance to its previous American-centered policy. On the other hand, the diversity—or disorganization—of European science policies gave the foundation a field of action that was both larger and freer than in the United States.

Some of the laboratories that received Rockefeller grants played an important role in the rise of molecular biology. In the 1930s and 1940s, Morgan's genetics laboratory at Caltech, like Pauling's and Beadle's, was heavily funded by the foundation. On the advice of the Canadian-born French scientist Louis Rapkine, the administrators of the Rockefeller Foundation made important grants to the Centre National de la Recherche Scientifique in France after the war. The funds were to be used both for equipping laboratories and for organizing international conferences. Among the key beneficiaries of this money were Boris Ephrussi's group and the laboratories of André Lwoff and Jacques Monod at the Pasteur Institute.[6] Chapter 14 describes the role of these French research groups in the development of a specific school of molecular biology.

By contrast, some groups that were to play a major role in the development of molecular biology in the 1950s and 1960s did not receive any money at all from the foundation. For example, Watson and Crick did not receive a Rockefeller grant for their work on the double helix, nor did the various members of the phage group receive any Rockefeller money. This is partly explained by the fact that, from the middle of the 1940s on, public funding for molecular biology rapidly increased, in particular through the National Institutes of Health (NIH) in the United States, the Medical Research Council (MRC) in Britain, and, a bit later, the CNRS in France.[7] This initial overview suggests that grants from the Rockefeller Foundation were more important for the prehistory of molecular biology, for the initial steps in its development, than for its subsequent progress.

Most historians of science give the Rockefeller Foundation an essential role in the paradigm shift that took place in biology in the 1940s. Only Pnina Abir-Am[8] has denied that the foundation had any positive role whatsoever, arguing that it awarded grants only to laboratories that were already relatively rich, to well-known research groups, and not to young groups or to truly innovative projects. Grants awarded to physicists even tended to isolate their recipients yet further, preventing them from adopting a genuinely biological approach—or at least not helping them to do so. The real molecular biology, argues Abir-Am, was born out of the close cooperation of biologists, biochemists, geneticists, and physicists, without any help from the Rockefeller Foundation. This criticism has led to a clarification of the exact role of the foundation, which can be summarized as follows.

The grant-awarding policy followed by Warren Weaver and the Rockefeller Foundation, which gave priority funding to physics and chemistry laboratories that wanted to start studying biology, or to biology laboratories to buy equipment that would enable them to make physical studies of organisms, was quite clearly based on a materialist, reductionist philosophy. According to this position, future progress in biology would come from an understanding of the molecules that constitute all life. Providing laboratories with the ability to use physico-chemical techniques undoubtedly played a key role in changing scientists' mentality. *A priori* biologists might think that studies of biological molecules would not be sufficient to understand the functioning of organisms. If, however, the availability of techniques and machines were to lead their research to the study of such molecules, there can be little doubt that their "philosophy" would be spontaneously shaken. Slowly but surely, they would be convinced that the physico-chemical study of biological molecules would be the foundation of the new biology. Too often, a "conceptual" presentation of the history of science makes it impossible to see just how much the models and the concepts of scientists are the prisoners of the technologies employed. Too often these techniques are little known or badly described in terms of the "mental" constraints they induce.

On a more concrete level, it can be shown that the laboratories supported by the Rockefeller Foundation played an essential role in the development of molecular biology, even if none of them, taken on their own, were at the forefront of the new science. Many of these laboratories pursued lines of research that turned out to be mistaken, but to highlight these errors, or the dead-ends they led to, would be to ignore the contingent aspect of scientific work. For example, Linus Pauling failed to discover the double helix structure of DNA, but this is of little importance compared with the essential role he played in the reduction of biological problems to the physico-chemical level. Surely the most important point is that in trying to discover the structure of DNA, Pauling, who already had a substantial reputation, showed how important this structure was for the whole of biology (see Chapter 11).

Pauling also played an important part in later developments in molecular biology and in the study of the role of genes in controlling protein structure. His reorientation toward biology was largely provoked by a substantial grant from the Rockefeller Foundation: in the space of approximately twenty years, Pauling's laboratory received nearly one million dollars from the foundation.

Pnina Abir-Am's final criticism of the role of the Rockefeller Foundation is the most serious: she accuses Warren Weaver of not having financed (and thus of having blocked) the creation of an Institute of Mathematical and Physico-Chemical Morphology at Cambridge University in 1935.[9] The project was intended to contribute to the study of embryonic development and morphogenesis through a molecular approach. It was supported by the crystallographer J. D. Bernal, by the theoretical biologist Joseph Woodger, by the embryologists Joseph Needham and Conrad Waddington, and by the mathematician Dorothy Wrinch.[10] According to Abir-Am, the lack of vision shown by Weaver and the other administrators of the Rockefeller Foundation meant that the opportunity represented by the molecular study of embryogenesis was simply allowed to slip away. The project was not completely rejected, however, as shown by the fact that Waddington was awarded a grant to visit U.S. genetics laboratories in 1937 and 1938. Furthermore, the studies carried out by this group of scientists were supported by the foundation, even if it did not encourage the creation of the institute.

Weaver had several reasons for rejecting the request to set up an institute. First, Cambridge University itself was not particularly enthusiastic about the project. Second, the British biologists Weaver consulted were also very reserved about both the nature of the project and the personalities involved, three of whom were well-known radicals. Finally, the approach they proposed (with the exception of Needham and Bernal) was the opposite of that of the foundation. Their approach was theoretical and antireductionist: they sought to use mathematical formulae to understand the biological problem of organization.

Waddington and Needham were well known for their previous work on the "organizer," which they saw as the starting point for the development of the Institute of Physico-Chemical Morphology. The problem of the organizer was and still is one of the most exciting and fundamental problems of development. The phenomenon of embryonic induction had been discovered as a result of experiments on early development in frogs and toads. As early as 1924, Hans Spemann's ablation and transplant experiments had shown that a part of the embryo (the mesoderm of the blastoporal lip)[11] could organize the development of the embryo and, for example, induce the development of neural cells in non-neural tissues. This phenomenon suggested that a substance of unknown physico-chemical nature was present in the mesoderm. Studies by Spemann's colleagues, followed by

Waddington and Needham, showed that induction was not altered when the mesoderm was treated with heat or alcohol and that cellular extracts were also active.

Studies aimed at characterizing the inducer—later rebaptized the evocator—present in these extracts all led to a dead-end: different components—glycogen, nucleic acids—were found to have an inductive power. Even completely synthetic compounds, such as methylene blue, were active. Finally, inductive activity could be extracted from tissues that showed no such effect.

The new tendency in historical research to break with the traditional history of science has led historians to concentrate on "failed" approaches and theories, putting them on an equal footing with successes.[12] Pnina Abir-Am's study of the project for the Institute of Mathematical and Physico-Chemical Morphology undoubtedly is a result of this tendency. Her argument is not entirely convincing for the following reasons:

- In May 1935, the project as presented was based upon research that was already headed in a problematic direction.
- The proposed approach was holistic, extremely nonreductionist. Waddington was very critical of molecular biology, while Woodger had already opposed the mechanistic and reductionist approach of genetics. It would, no doubt, have been highly commendable for Weaver to have funded a group of researchers with an ideology so far removed from his own. But given the people involved, it is unlikely that such an institute would have contributed to the development of molecular biology.
- It is difficult not to look retrospectively at the proposed project: until very recently, the problem of embryonic induction remained unresolved. As J. Slack put it, the search for the compounds involved in induction was a bit like "looking for a contact lens in a swimming pool, with the added possibility that the contact lens might be soluble in water."[13] There are a number of problems in science that are theoretically important but that experience shows should be avoided. No scientist would ever say so out loud, but nevertheless, avoiding this kind of project is part and parcel of the unwritten training of all scientists.

- The idea of a mathematical approach to the structure and functioning of organisms also turned out not to be a good choice. Dorothy Wrinch proposed a model of protein structure (the cyclol theory) that was chemically incorrect and was brusquely rejected by Pauling[14] (see Chapter 9).

In sum, the Rockefeller Foundation certainly played an important role in the development of a molecular approach to biology that was based on the use of the techniques of chemistry and physics. This led to the more or less unconscious validation of a reductionist conception of biology. Yet the foundation should not be presented as the driving force behind this evolution: by its grant-awarding policy, it was both the reflection and the servant of a more general movement within science (described in Chapter 7).[15] This movement, which was not restricted to a single institution or a single country,[16] also led to the birth and growth of biotechnology.[17]

CHAPTER 9

A New World View

The best way to understand the nature of molecular biology is not to read popular accounts but, quite simply, to visit a molecular biology laboratory. There is a striking consistency in the techniques and apparatuses that are used. Some of these have been altered by the genetic engineering revolution, but the fundamental principles remain the same.

All molecular biology labs employ bacteriological techniques, some of which go back many decades. The culture media, the form of the instruments, and even the way in which the equipment is handled were all established at the end of the nineteenth century by the French and German schools of bacteriology (Louis Pasteur and Robert Koch, respectively).[1] The phage group—Delbrück, Luria, and their colleagues—adopted and adapted these techniques.

There are also techniques for separating biological macromolecules and their constituents. These methods have been continually improved throughout the twentieth century, and even though these changes have not always been spectacular, they have accompanied theoretical advances. For the first half of the century, the most widely used method for fractionating proteins was salt precipitation. It was slowly replaced by chromatography.

In chromatography (long used in organic chemistry and, more recently, in biochemistry) a liquid phase is passed through an absorbent, porous, and rigid material. The development of "partition chromatography" and of new detection methods by the British scientists A. J. P. Martin and R. L. M. Synge rendered this technique sufficiently sensitive to make possible the analysis of the constituents of biological macromolecules, and in particular the separation of amino acids.[2] Instead of using a normal solid adsorbent material, Martin and Synge used an inert powder as the material support for the liquid phase. The column was washed with a second liquid phase that was immiscible with the first. In this kind of chromatogram, the partition between the two liquid phases replaces the adsorption on a solid phase. The material support used for the first phase was initially silica, then paper.[3] Another British scientist, Frederick Sanger, immediately used this new technique to try to determine the amino acid sequence of the insulin molecule (see Chapter 12). Intrigued by Avery's results, Erwin Chargaff used the same method when he tried to test Levene's tetranucleotide theory and to repeat the measurement of the composition in bases of different nucleic acids[4] (Chapter 3).

Various other chromatographic techniques were developed or adapted for the separation of biological macromolecules: chromatography on ion-exchanging resins[5] or dextran beads that separated molecules according to their size.[6] The most recent of these techniques is affinity chromatography: a molecule that has a high affinity with the protein to be purified is fixed to an inert material support.[7] The subsequent application of high pressure makes it possible both to speed up the purification steps and to increase their effectiveness by restricting diffusion.

But the two techniques that best characterize molecular biology are ultracentrifugation and electrophoresis. These techniques can be used with two rather different aims: an analytical objective—the characterization of the properties of biological macromolecules (mass, form, electric charge, and so on); or a preparative objective—the purification of these macromolecules, their separation from contaminating molecules. The development and diffusion of these techniques in biological laboratories were greatly assisted by the Rockefeller Foundation. These techniques changed the work and objectives of biologists and encouraged the adoption of a molecular vision of biology. They are sufficiently important for their history to be dealt with in some detail.

Rarely has a technique (ultracentrifugation) been so closely linked to a place (the University of Uppsala) and to a researcher (Theodor, or The, Svedberg).[8] The first centrifuge used to determine the molecular mass and the properties of proteins was developed in 1924, and the first measurements were made in 1926.

As noted, it took some time for the concept of the macromolecule to overcome the influence of colloid theory. According to colloid theory, biological substances were composites formed by the aggregation of small molecules; their molecular mass would thus be the mean of the mass of the different aggregates. According to the macromolecular approach, however, molecular mass had a precise value. Measuring this mass was thus crucial. Existing techniques, which measured the diffusion of light, or variations in osmotic pressure or viscosity, gave only imprecise results.

The first experiments aimed at deducing molecular mass from the speed of sedimentation during rapid centrifugation had been carried out at the beginning of the century. The (slow) speeds obtained had permitted only the mass of microbeads to be measured.

Theodor Svedberg,[9] trained in the physico-chemical study of colloids, had initially studied Brownian motion; he became interested in ultracentrifugation only in the 1920s. During a visit to the University of Wisconsin in 1923, he developed optical methods for observing the sedimentation of proteins during ultracentrifugation. On his return to Sweden, he built a machine that could produce a centrifugal force 7,000 times that of gravity (*g*). Two years later, he finished a new machine that could produce a centrifugal force of 100,000 *g*. With this apparatus, he was able to determine the molecular mass of hemoglobin—the protein that transports oxygen in the blood—and to show that it behaved like a macromolecule with a precise molecular mass. In the years that followed, Svedberg measured the molecular mass of a series of purified proteins, hormones, and serum components. Each time, exact values were obtained. Stanley sent Svedberg recently crystallized samples of the Tobacco Mosaic Virus so that he could measure their molecular mass (see Chapter 6). Indeed, Svedberg selected the proteins he studied as a function of the work being carried out at the Rockefeller Institute.

The history of science is never simple. These initial results of ultracentrifugation led to the definitive decline of the colloid theory and the triumph of Staudinger's macromolecular theory. But they also seemed to indicate that protein molecular masses were relatively consistent. Several of the proteins studied by Svedberg turned out to be made up of sub-units with a constant molecular mass of around 35,000 daltons (a value that was subsequently reduced to 16,500 daltons[10]). This value took on great significance and led to the development of a number of models of protein structure. On the basis of these data, Dorothy Wrinch,[11] a member of the Cambridge Theoretical Biology Club who worked with C. H. Waddington and Joseph Needham on the Institute of Mathematical and Physico-Chemical Morphology project, proposed the cyclol theory. According to this theory, protein structure was stabilized by covalent bonds other than the peptide bond. In a 1939 paper published in *Science,* Linus Pauling and Carl Niemann rejected this model and argued that if proteins have preferential molecular weights, this could not be explained at the chemical level but had a biological origin that was linked to the evolution of life.[12]

The supposed existence of privileged molecular masses turned out to be an artifact produced by the fact that very few proteins had been studied. The interest stimulated by these preliminary results showed that many biochemists hoped to discover simple rules that would explain the structure of proteins, and above all their formation (see Chapter 12).

Svedberg thus played an essential role in the development of analytical ultracentrifugation. In 1926 he received the Nobel Prize—not for this work but for his studies of Brownian motion, carried out twenty years earlier. The theoretical value of these studies had been strongly criticized by Einstein; Jean Perrin would later attack their experimental value. Svedberg found the circumstances surrounding his Nobel Prize embarrassing: during initial discussions in the special commission of the Swedish Academy of Sciences, he refused to allow his name to go forward for the prize. In a subsequent plenary meeting, however, he kept silent, to the great surprise of the other members of the commission. As he wrote in a recently discovered autobiographical note, "I promised myself to use the following ten years of my life to make myself worthy of the prize."[13] Although the Nobel Prize is usually awarded for a major scientific achievement, in Svedberg's case the award precipitated the achievement. It made his work easier and enabled him to acquire the equipment necessary for his subsequent experiments. The Swedish parliament gave more than one million crowns toward the construction of a Department of Physical Chemistry, to which, in 1926, the Rockefeller Foundation added $50,000 to buy equipment.

The links between the Rockefeller Foundation and The Svedberg were established early on, at a time when support for groups studying biological problems with physical tools was not yet the foundation's main priority. These links were maintained over the years that followed. In 1936, the University of Uppsala received more than $250,000 from the Rockefeller Foundation. Warren Weaver was a personal friend of Svedberg's (they met in 1923 at the University of Wisconsin). This explains how the Rockefeller Institute at Princeton, where William Stanley and John Northrop worked, rapidly acquired an ultracentrifuge as powerful as that developed in Sweden. This centrifuge, acquired with Svedberg's assistance, helped the institute to take a leading position in viral research and to maintain this position in the years that followed (see Chapter 6).[14]

Ultracentrifuges are not only important for measuring molecular mass, but can also be used for the preparation of samples—for separating different macromolecules. Beginning in the late 1920s, relatively slow "preparative" centrifuges were widely used in biochemistry laboratories. High-speed preparative centrifuges—ultracentrifuges—became available only later. Despite being simpler than analytical ultracentrifuges (they did not need the same optical equipment for taking measurements during centrifugation), they were nevertheless subject to the same mechanical constraints.

Preparative centrifuges are still used by protein biochemists and molecular biologists. Their uses are varied, depending on the speeds they can attain and, thus, the centrifugal forces they can produce. Low-speed machines make possible the collection of a raw extract following the rupture of the cells in the sample and the isolation of proteins or nucleic acids following precipitation by the addition of salts or alcohol. High-speed machines make it possible to isolate viruses and separate the various cellular structures ("organelles"), each of which plays a particular role in the life of the cell—energy production, protein synthesis, and so on.[15] With a few experimental tricks, ultracentrifuges even enable scientists to separate proteins. Fractionation by centrifugation and ultracentrifugation of intracellular contents led to the isolation of cellular extracts in which the first *in vitro* protein synthesis was carried out, and on the basis of which the genetic code was deciphered (see Chapter 12).

Jesse W. Beams and Edward G. Pickels of the University of Virginia made several essential modifications to the ultracentrifuge that made it reliable and easy to use: they put the centrifugation chamber in a vacuum to avoid over-heating, then used an electric motor to drive the centrifuge.[16] In the middle of the 1940s, Pickels set up a company called Spinco (SPecialized INstruments COrporation), which, in less than a decade, sold more than three hundred models of its new ultracentrifuge. The history of the installation of these machines in biochemistry and molecular biology laboratories has yet to be written: it will, no doubt, teach us a great deal about the diffusion of theoretical models and concepts in both disciplines.

Like ultracentrifugation, electrophoresis had a wide range of functions.[17] During the early years, it was used as an analytical method for characterizing macromolecules—in particular proteins, but also nucleic acids—to show that these compounds corresponded to unique substances with well-defined properties. It was also used as a preparative technique, in the separation of both proteins and their fragments. With the development of genetic engineering, electrophoresis took on new importance; today it is *the* technique used to separate DNA fragments and to determine the sequence of bases in DNA molecules (see Chapter 16).

The idea of using an electric field to separate different biological components is not new. Once again, the first practical application was at least partly the brainchild of The Svedberg, who, after a number of failures, handed the project to his student Arne Tiselius. In 1948 Tiselius received the Nobel Prize for his development of the apparatus.

The first results were disappointing. Like other researchers who had tackled the problem, Tiselius had difficulties with heating and diffusion, so much so that he more or less abandoned the project until his 1934 visit to Princeton University on a Rockefeller Foundation grant. During his stay he met many U.S. protein researchers as well as the Rockefeller Institute immunochemists—Wendell Stanley, Karl Landsteiner, Leonore Michaelis, and Michael Heidelberger—all of whom emphasized the importance of developing a new method for separating and characterizing proteins. He returned to the project, concentrating on eliminating those sources of variation that prevented electrophoresis from being a precision tool. A new, extremely large apparatus was completed in 1936 with the aid of a Rockefeller Foundation grant. The problems of heating and diffusion were resolved. The optical system employed in the ultracentrifuges was used, so that direct observations could be made of the different components in the sample to be analyzed. It was even possible to remove parts of the sample during electrophoresis, thus making the apparatus both an analytical and a preparative tool.

Tiselius's first results with the new apparatus were spectacular: he was able to separate the different proteins present in serum. Together with Elvin Kabat and Michael Heidelberger, he showed that the antibodies were found in the γ-globulin fraction. Tiselius's apparatus was subsequently copied and improved, but even in its initial version it enabled L. G. Longworth at the Rockefeller Institute to show that there were differences in the serum protein composition of healthy and sick subjects, thus opening the road to medical applications for the new technology.

Nevertheless, the electrophoresis apparatus was cumbersome, expensive, and difficult to use, requiring great skill. By 1939, only fourteen such machines were in use in the United States, five of them in Rockefeller Institute laboratories. Most of these machines had been bought with Rockefeller Foundation money. In 1945, the American company Klett made the first commercial electrophoresis machine. By 1950 four different companies were making them, and selling them at a considerably reduced cost.

But it was only in the early 1950s, when optical observation was replaced by easily detectable labeled molecules, that the use of electrophoresis became truly "democratic."

In addition to liquid phase electrophoresis, the technique was also carried out on a silica gel (or paper).[18] From the 1940s this method was widely used to separate amino acids and nucleotides. It was eventually replaced by electrophoresis on agar (a solid sugar polymer made from seaweed), then starch, which in turn was replaced by the polyacrylamide gel, which, from the 1960s, became the main technique used in molecular biology for separating macromolecules.[19]

It took many years for electrophoresis to have the same impact on the development of molecular biology as ultracentrifugation had. Nevertheless, according to the U.S. historian Lily Kay, electrophoresis led to a situation where "vital processes . . . were increasingly probed through systematic applications of tools from the physical sciences. This trend altered the nature of biological knowledge, the organization of research"[20] (in particular by involving more and more physicists in biological research).

Physics also made a decisive contribution to the new biology through the creation of "labeled" molecules—molecules that behaved chemically like normal molecules but could be distinguished by physical criteria (mass or radioactivity). Detecting these differences makes it relatively easy to follow the fate of these molecules, both in the test tube and *in vivo*. One of the differences between Avery's experiment and the Hershey-Chase study was that Hershey and Chase used radio-labeling to distinguish DNA from proteins. This new technology increased the impact of their experiment.

George Hevesy was one of the first scientists to use isotopes when, in 1923, he measured the uptake of a lead isotope by plant roots. As we have already seen, Hevesy joined Niels Bohr's group in Copenhagen to continue his work on the use of isotopes in biology and medicine.

In fact, most of the molecules that were initially labeled were not radioactive, but contained heavy isotopes of certain elements.[21] Greater sensitivity of detection techniques meant that radioactive isotopes later almost completely replaced heavy isotopes, to the extent that it has been completely forgotten that for ten years or so heavy isotopes made it possible for biology to make considerable advances. This forgotten chapter in the history of biology is worth studying in detail.

In 1931, in the Chemistry Department of Columbia University, Harold Urey discovered deuterium, a heavy isotope of hydrogen. This discovery quickly led to the use of isotopes in biological research. The first description of the use of "deuterated" compounds for the study of intermediate metabolism was published in 1935 by Rudolph Schoenheimer and David Rittenberg of the Department of Biochemistry in the College of Physicians and Surgeons at Columbia University. In the same year, the Rockefeller Foundation set up a special fund intended for chemists who knew how to handle deuterium and who wanted to incorporate it into biologically interesting molecules. These studies developed on an even larger scale when Urey obtained compounds enriched by the heavy isotope of nitrogen, ^{15}N, which were immediately used by Schoenheimer and his coworkers. Deuterated compounds were well adapted to the study of lipid metabolism, whereas ^{15}N was particularly well suited to the study of amino acid and protein metabolism. But though deuterium could be relatively easily detected (it simply required a few measures of density or refraction), the study of molecules containing ^{15}N necessitated far more complex equipment—a kind of mini particle accelerator called a mass spectrometer. This restricted the use of this isotope to a handful of well-equipped laboratories.

Schoenheimer extended his studies to proteins and then to all biochemical substances. After working with Hevesy, he was forced in 1933 to flee Germany for the United States, where he obtained a position in the Department of Biochemistry at Columbia University.

Schoenheimer's research convinced him that all the constituents of the organism were in a state of dynamic instability. His conclusions were described in an important book published in 1942 after his death, *The Dynamic State of Body Constituents:* "The large and complex molecules and their component units, fatty acids, amino acids, and nucleic acids, are constantly involved in rapid chemical reactions . . . The free amino acids are deaminated, and the nitrogen liberated is transferred to other previously deaminated molecules to form new amino acids. Part of the pool of newly formed small molecules constantly re-enters vacant places in the large molecules to restore the fats, the proteins and the nucleoproteins."[22]

The conclusions reached by Schoenheimer and his colleagues had an important impact on biochemistry, in particular on the analysis of metabolism. Nevertheless, the influence of Schoenheimer's book was ambiguous: by arguing that proteins had the same metabolic instability as amino acids, and by suggesting that the amino acids that form proteins could be continually added and subtracted, Schoenheimer conflated the biosynthetic and degradation pathways of proteins with those of amino acids. He thus placed the problem of protein synthesis in the context of metabolism in general. His solution was the opposite of that which molecular biology was to propose (see Chapter 12). Twenty years later, the French molecular biologist Jacques Monod acknowledged that Schoenheimer's book had an important influence, and that he and other researchers had found it hard to believe that, on the basis of the results obtained for β-galactosidase, proteins were synthesized in a single step, and once synthesized were metabolically stable (see Chapter 14).[23]

The research of Schoenheimer, Urey, and others was largely financed by the Rockefeller Foundation, which, among other things, helped them to buy mass spectrometers. Warren Weaver directly oversaw the development of their studies.

From the beginning of the 1940s radioactive isotopes gradually replaced heavy isotopes for labeling molecules. The first radio-isotopes produced by particle accelerators had a very short half-life and thus did little to affect the virtual monopoly of the heavy isotopes. Only radioactive phosphorus, as used by George Hevesy, was an important tool for biochemistry. In February 1940, in the Lawrence Laboratory at Berkeley, Martin Kamen and Samuel Ruben discovered the radio-isotope ^{14}C, which had a very long half-life. This heralded the end of the golden age of heavy isotopes.[24]

The use of heavy isotopes involved a combination of skills in physics, physical chemistry, and chemistry. The use of radio-isotopes, on the contrary, merely required the acquisition of a Geiger counter: their use spread rapidly between 1940 and 1950.

Ultracentrifugation, electrophoresis, and isotopes all show the same process of transformation: initially extremely cumbersome and requiring the close collaboration of physicists, chemists, and biologists, these techniques were gradually simplified enough to be employed in all biology laboratories.

The same process can be observed in the use of other physical techniques such as spectroscopy,[25] the study of the absorption of light—visible or ultra-violet—by molecules. In this case, too, the Rockefeller Foundation played a very important role in developing the first apparatuses and in funding experiments that showed the importance of this new technique in the physico-chemical study of biological phenomena.[26]

One final technique from physics that contributed to the growth of molecular biology was the electron microscope,[27] which made a direct contribution to a major discovery—the structure of the bacteriophage (Chapter 4). Apart from this example, however, the electron microscope played a secondary role, even in the identification of intracellular structures (Chapter 13), despite what a naively realist conception of science might imply. The images produced by the electron microscope were in fact too complex: their correct interpretation required a pre-existing model. Furthermore, the energy from the electron beam was so great and the chemical treatments involved in "fixing" (stabilizing) the sample and thus increasing the contrast so extreme that the technique produced a large number of artifacts and made pseudo-structures appear where none was present.[28] Images produced by electron microscopy no doubt had a certain pedagogic value and were useful as documentary proof, but they did not directly lead to any discoveries. Electron microscopy simply showed the reality of results that had generally been obtained using other techniques.

CHAPTER 10

The Role of Physics

Physicists and technicians from the world of physics played a fundamental role in the birth of molecular biology, but historians of the subject disagree as to the nature and extent of this influence. H. F. Judson, for example, has argued that the concept of information, thought by many to be inherited from physics, is virtually indispensable for explaining the main discoveries of molecular biology. But though this concept is of fundamental importance today, it was not so at the time: it played no role in the development of molecular biology.[1]

For Arthur Kornberg, physicists who turned to biology introduced many of the strategies of physics into their new field, but few of its tactics.[2] In other words, they changed the form of biological research without altering its fundamentals. For Alain Prochiantz, molecular biology is simply an avatar of biology. Schrödinger in no way favored the growth of molecular biology (it existed long before *What Is Life?*)—the interest of the book is to be found in the "analogies" it suggested.[3]

The bitterness of many of these discussions reflects the fact that the stakes go far beyond the history of molecular biology, relating to the orientation of current research and its financing. Short of a definitive answer to this debate, a few remarks can be made:

- If physicists did influence the development of biology, biology was nevertheless not reduced to either chemistry or physics. There is no trace of any mathematical or physical formalism in contemporary biology.
- One of the arguments put forward to refute the idea that physicists had any real influence is that biological research shows an apparent continuity and that there was no sudden change in orientation during the crucial period of the birth of molecular biology. This continuity, however, does not imply that there was not a "molecular revolution"—a scientific revolution always takes place against a backdrop of stability. Only such continuity in research and in the subject being studied makes possible the confrontation of old and new scientific visions.[4]
- Both those who accept and those who reject the influence of physicists on the development of molecular biology generally share the same mistaken vision of scientific progress. In order to find the key influences on a science, they look to its origins, to its infancy—this is a kind of "preformationist" view of the history of science. H. F. Judson's comments are particularly revealing from this point of view, because they underline that it is impossible to detect any influence of the concepts of physics on the early experiments of molecular biology. And yet these concepts have become indispensable for describing the discipline. This paradox can be explained simply by the fact that today's molecular biology was not present in the initial experiments that founded the science, and that it has only gradually taken the form we know today.
- Those concepts that are so useful for understanding and explaining the results of biological research (information, feedback, programs, and so on) have not been "borrowed" from physics (or from computing). Instead, they are common both to these disciplines and to biology and can be found at the heart of the new world view that developed during and after the Second World War. They constitute the framework within which the experiments, theories, and models of molecular biology took shape.
- The migration of scientists from Europe to Britain and, in particular, to the United States, as a consequence of the rise of fascist and Nazi regimes, played an important role in the birth of a new world view. The forced confrontation of different scientific traditions, the forging of links between scientists working in very different domains but sharing the same difficult exile status, were important aspects of a rich cultural mix.[5]

- The development of the techniques of genetic engineering shows that the molecular understanding of biology, acquired between 1940 and 1965, was an *operational* understanding.[6] Today both molecular biologists and physicists share a scientific world view in which knowledge and action are intimately linked. Physicists played an important role in this change in the form of biological knowledge, by the way they conceived and carried out their experiments. In following Delbrück and asking simple questions of biological objects, they obliged these objects to reply in the same simple language.
- The most important contribution of the physicists was perhaps simply to have been convinced (with a certain dose of naiveté) and to have convinced the biologists that the secret of life was not an eternal mystery, but was within reach.

Historians of science have a great deal of work to do before we can hope to understand what happened in the period from the 1940s to the 1960s, when two fundamental disciplines—computer science and molecular biology[7]—were born and developed, when a new world view appeared according to which information and logic count more than energy or material constitution.[8]

PART II

The Development of Molecular Biology

CHAPTER 11

The Discovery of the Double Helix

Few scientific findings are as well known as that of Jim Watson and Francis Crick, who, in 1953, discovered the double helix structure of DNA.[1] The discovery was not the work of a group or a laboratory, but a result of the meeting of two extraordinary personalities who had been trained in different scientific disciplines but had complementary abilities. Crick, a physicist by training,[2] had begun his research at University College, London, where he studied the viscosity of water heated at high pressure. Having worked for the Admiralty on the development of magnetic mines during the war, he decided after 1945 to turn to biology. Recruited by the Medical Research Council (MRC) to study the movement of small magnetic particles within cells, he heard about a new research group being set up at the Cavendish Laboratory at Cambridge University. This small team, led by Max Perutz and supervised by Sir Lawrence Bragg, was working on the description of protein structures by X-ray diffraction.[3] Crick asked to be transferred to this group, and got his wish.

The protein Perutz decided to study was hemoglobin. The work involved purifying it, obtaining crystals, then directing a beam of X-rays at the crystals. The diffraction of the beam and its decomposition into a number of different beams left a trace on a photographic plate placed behind the crystals, forming a diffraction pattern. The theory of diffraction, developed by Lawrence Bragg and his father, explained that the distribution and intensity of the diffraction pattern were a consequence of the structure of the molecules present in the crystal.

Research carried out before the war by Bragg's group in England and by Linus Pauling in California had shown that, by using crystals of small molecules, it was possible to deduce the three-dimensional structure of the molecules from the diffraction pattern. Bragg's[4] aim was to study more complex molecules and to show that the diffraction technique could help determine the structure of molecules as complex as proteins. This was not an easy task. In the first place, the photographic diffraction images were not very good. There were also a number of major theoretical problems. The size of the diffraction patterns was the result of the diffraction of a wave with a given intensity and a given phase. But the physicist had access to only one value in determining the two factors. To these theoretical problems was added the fact that the calculations required to interpret the images were extremely complex and had to be carried out with the aid of calculators (computers had not yet arrived in crystallography laboratories).

Protein crystallographers were thus unable to interpret the diffraction images—they could propose only vague models of the form of the proteins they were studying, theoretically determine the diffraction spectrum of these models, and see if the theoretical results were compatible with the observed images.

One of the characteristics of a diffraction image, or of some of the curves that can be derived from an image, such as a Patterson curve, is that they tend to be symmetrical and regular. The symmetry is a reflection of repetitions and regularities in the crystals and in the molecules that form them.

Francis Crick became familiar with the complexity of analyzing diffraction images and quickly realized the limitations of the tools available. His generally negative attitude was not appreciated by his laboratory colleagues.[5] A thirty-something who still had not finished his thesis, he was competent but somewhat skeptical.

James D. Watson had been a brilliant student. Having completed his studies in biology at an early age, he was Luria's first student. For his doctoral research he had studied the effect of X-rays on the development of bacteriophages. The data were difficult to interpret. Luria and Watson were convinced that to understand these results and, more generally, to understand experiments on the bacteriophage, it was necessary to have a better understanding of the chemical constituents of the phage, and in particular of DNA. The phage group increasingly suspected that DNA played an essential role in phage replication: the end of Watson's doctoral studies preceded by a mere two years the publication of the Hershey-Chase experiment (see Chapter 4).

Luria and Delbrück thought that European science was more imaginative than its American counterpart and that it would be ideal for a brilliant young researcher like Watson to do his postdoctoral research in Europe. Luria decided to send Watson to Copenhagen to work with Herman Kalckar, a specialist in amino acid metabolism whom he had met at the Cold Spring Harbor phage course.

Watson's visit to Kalckar's laboratory turned out to be relatively unproductive, however, apart from a study on phage that he carried out with Ole Maaløe in another laboratory in Copenhagen. While on a visit with Kalckar to the Naples Zoological Station, Watson participated, almost by accident, in a conference on macromolecules. He was struck by Maurice Wilkins's talk on the diffraction of X-rays by DNA fibers and decided to prolong his stay in Europe in order to work in a crystallography laboratory. Luria contacted John Kendrew, who worked with Max Perutz, and, with Perutz's agreement, Watson joined the team at the Cavendish Laboratory.

Watson's official project was to purify and crystallize a protein that was analogous to hemoglobin, but simpler—myoglobin. This project never came to term. As soon as he arrived at Cambridge, Watson began a close collaboration with Crick aimed at determining the structure of DNA. For Watson and Crick, understanding this structure was fundamental for understanding the role of genes.

At first glance, Crick and Watson did not have very much going for them. Although their training was complementary—Crick could provide

Watson with the information necessary for understanding the principles and interpreting the results of crystallography, whereas Watson could inform Crick about developments in bacterial genetics and the latest results from the phage group—their handicaps were far greater. Crick had to spend an important amount of time completing his doctoral thesis, while Watson had to devote time to crystallizing myoglobin, the project for which he had been recruited. But the biggest obstacle was that neither of them was able to determine the diffraction spectra of DNA fibers—they had to rely on results previously published by William Astbury, and especially on data provided by Maurice Wilkins of King's College, London.

During the war Maurice Wilkins—who, like Crick, had been trained as a physicist—had worked at Berkeley on the separation of uranium isotopes as part of the Manhattan Project. Like Crick, he had turned to biology after the war. After working on ultra-violet microscopy, he began studying DNA fibers by X-ray diffraction. Using high-quality DNA samples, he was able to obtain very precise diffraction images and to show that the degree of hydration played an important role in the structure of DNA molecules.

Wilkins had recently been joined by Rosalind Franklin,[6] a physical chemist by training who had an excellent knowledge of X-ray diffraction, which she had previously used in a study of the structure of carbons. Franklin was an exceptional experimentalist: she rapidly produced images that were far better than any previously obtained.

Wilkins's group thus apparently had all that was required to make rapid progress in the study of DNA structure. There was a fundamental problem between Franklin and Wilkins, however: when Franklin arrived at the laboratory she thought she had her own research project; Wilkins, by contrast, thought she was there simply to provide the technical expertise necessary for him to carry out his research successfully. This major misunderstanding limited the exchanges between members of the group and prevented the elaboration of a clear research strategy.

On the basis of data presented by Wilkins in informal discussions and at scientific conferences, Watson and Crick began to build models of the DNA molecule. They had very little to work on: the proportion of water present in DNA was not known, and the structure of the bases was still a matter of debate. The only thing everyone agreed on was the structure of the polynucleotide chains.

Linus Pauling was to play a decisive role—directly and indirectly—in the development of Crick and Watson's work. In a series of articles published in 1950 and 1951, Pauling proposed the structure of some elementary "motifs" that were present in proteins.[7] The first images of X-ray diffraction by proteins had shown that they contained repeated regular structures. As early as 1933 the British crystallographer Astbury, working at the University of Leeds, had proposed models for these structures on the basis of his study of the proteins that form wool and silk—α- and β-keratin.[8] More recently, models of polypeptide chains with a regular helical conformation had been presented by Lawrence Bragg, Max Perutz, and John Kendrew.[9] Pauling completed and corrected these initial attempts. The superiority of Pauling's work was a result of his excellent knowledge of the stereochemistry of the peptide bond and of hydrogen bonds that could play a role in stabilizing protein conformation.[10] This knowledge came both from a simple use of quantum mechanics and from the direct measurement of angles and bonds by X-ray diffraction on crystals of small, "model" molecules. If structural constraints were respected, there was only a limited range of possible conformations for polypeptide chains. By using simple but precise molecular models, Pauling proposed the existence of regular structures that were soon found to exist in proteins.

Pauling showed that the helix was an important structure in macromolecules. This idea was already widely accepted by crystallographers and by specialists in the structure of macromolecules; Pauling's results thus merely confirmed something they were already convinced of.

More significant was the lesson of Pauling's approach: valid predictions could be made only if the stereochemical constraints of the molecules were rigorously respected. But Pauling's most important contributions were perhaps his decision to turn to the study of the structure of DNA and his publication of an initial model that turned out to be completely wrong.[11] For Watson and Crick, Pauling's interest confirmed the importance of their work and stimulated their research. They first tried different models of double and triple helixes in which the bases were situated on the outside of the molecule. Then, partly in desperation, but perhaps also influenced by the work of John Gulland, who, as early as 1947, had suggested that DNA bases interacted through hydrogen bonds,[12] Watson opted for the opposite model, in which the bases were on the inside of the molecule. Thanks to the work of the chemist June Broomhead, Watson had a thorough understanding of the hydrogen bonds that could be formed between the bases.

Watson made a first double helix model of DNA in which the bases paired with themselves—adenine with adenine, thymine with thymine, and so on. This model immediately suggested how the strands of the DNA molecule could replicate themselves. This model was in turn abandoned when Jeremy Donohue, who had worked with Pauling for several years, pointed out to Watson that he had chosen the wrong tautomeric forms for the bases, that is, he had put the hydrogen atoms in the wrong position on the bases. Having abandoned both the "external" model and the pairing of identical bases, Watson had the idea of pairing different bases. He discovered that the base-pairs A-T and G-C had the same spatial structure, thus enabling the construction of a perfectly regular double helix. He realized that the pairing of A with T and G with C explained the existence of equal concentrations of A and T and of G and C, which had been discovered by Chargaff and which until then had been inexplicable (see Chapter 3).

Watson and Crick quickly wrote an article, which they sent to *Nature.* It was published on April 25, 1953, together with an article by Maurice Wilkins, Alexander Stokes, and H. R. Wilson, and another by Rosalind Franklin and Raymond Gosling.[13] These last two articles contained X-ray diffraction data, some of which were already known to Watson and Crick (Perutz had given them a copy of a confidential report sent to the MRC by Sir John Randall, Wilkins's superior), and others of which they were unaware and which confirmed the validity of their model.

They then sent a second article to *Nature,* this time dealing with the genetic implications of the double helix structure of DNA. This article was published on May 30, 1953,[14] and showed how, on the basis of this structure, it was possible to propose a very simple model for the replication of

DNA, and thus "explain" the auto-replicative power of genes. The two strands of DNA could separate and the bases on each strand would associate with the complementary bases present in the cellular medium in the form of free nucleotides. "All" that would be required would be for the nucleotides to bond together to produce two daughter DNA molecules, each of which would be identical to the parent molecule. Watson and Crick also suggested that mutations resulted from rare transient forms of bases that were responsible for incorrect pairing during replication.

Watson presented this structure, and its genetic implications, to the XVIIIth Symposium on Quantitative Biology, which took place at Cold Spring Harbor in the summer of 1953. Max Delbrück had widely publicized Watson and Crick's results by giving photocopies of their *Nature* articles to the participants in the symposium. Despite some criticisms, the double helix structure of DNA was rapidly accepted by the scientific community. A long series of visitors came to admire the model at the Cavendish Laboratory. A page had been turned in the history of the young science of molecular biology. The structure of DNA, of the gene, had been discovered, and this structure explained the auto-replicative properties that had so fascinated biologists since the beginning of the twentieth century.

It is difficult for the historian to know how to deal with Watson and Crick's discovery. It appears that everything that could be said about it has been said, that there is no unknown element that could falsify this or that aspect of the story, or at the very least provide a new and original point of view. Paradoxically, this information overload weakens our understanding. The "total" transparency of the discovery of the double helix erases the very object of historical research: nothing remains to be explained because everything is already known.

In fact, a discovery as well known as that of the double helix presents an open field to the historian of molecular biology: using details and anecdotes, it is possible to recreate the scientific context and the social forces that were at work in the development of the new discipline.

The scientific careers and personalities of the codiscoverers of the double helix are particularly revealing. Francis Crick and Maurice Wilkins were members of the group of physicists who turned toward biology in the anticipation of major new discoveries. They discovered biology through genetics, or rather through the distorted image of genetics presented by Schrödinger in *What Is Life?* Their change of direction was encouraged by research administration policy decisions: the MRC had set aside a number of positions for physicists.

In Britain, an obvious area of research for such scientists was the study of the structure of macromolecules by X-ray diffraction. This method had been developed by William Bragg, Lawrence Bragg's father, and had grown as a result of the work of his son and other researchers, such as J. D. Bernal (first at Cambridge, then at London) and William Astbury (at Leeds). Bernal was the first crystallographer to obtain clear images of X-ray diffraction by proteins. To do this, he had placed the protein crystals in a capillary tube that was closed at both ends, thus avoiding dehydration during the experiment. Although the first images, obtained in 1934, did not lead to a three-dimensional description of protein structure,[15] they were nevertheless perfectly clear, confirming the macromolecular nature of proteins (to the extent that this was still necessary). Resolving the three-dimensional structure thus became a "do-able" project.

The British school of crystallography was without doubt the best in the world. But this strength was also its weakness. The crystallographers were convinced of their ultimate success: continued progress in X-ray diffraction showed that the determination of the structure of biological macromolecules—DNA and proteins—was quite possible. But armed with this certainty, and unaware of the importance of these structures for molecular biologists, for whom the chemical nature and structure of genes had become a key problem, the crystallographers were in no particular hurry. Crick and Watson thought that Wilkins and Franklin worked very carefully but far too slowly, because the structure of DNA was only of minor importance for them—a small step in the long march of scientific knowledge.

Watson's impatience was representative of the change of attitude in the members of the phage group. In the initial phases of his research, Delbrück had hoped to explain the self-replication of genes and bacteriophages without having to understand their chemical composition. All the research carried out in this initial period had shown that the phage was extremely complex and that it was hopeless to expect to understand how it replicated without opening the "black box"—without trying to understand the nature and the structure of its component parts. Irradiating bacteria during the replication of the phage, as Watson had done during his doctoral studies, was not enough to reveal the mechanisms of replication.

Luria was well aware of this, so he encouraged Watson to work on the biochemistry of nucleic acids. This work, however, contained a trap that Watson only narrowly escaped thanks to a series of lucky chances; this danger was to become embroiled in the complexity of the metabolism of nucleotides and bases, to get lost in the pathways of biosynthesis and degradation.

Watson and Crick's discovery presents us with an analytical problem similar to that encountered earlier with Avery's work. Both Avery and his readers were astonished by the result he obtained, because they were convinced that DNA was not an important, specific molecule. On reading Watson and Crick's articles and studying the reactions to them, the surprise is exactly the opposite. By this stage, everyone was convinced of the role played by DNA in heredity and no one felt it necessary to justify this conviction. Several theories have been suggested that might explain the change in attitude that took place in molecular biology, this new conviction that DNA was fundamental to heredity. These theories nevertheless do not fully explain a shift in opinion that resembled a conversion.

We should reject an idea that is widely held by both the admirers and the detractors of molecular biology. According to this explanation, Watson and Crick inaugurated a new style of research in which discussions and theories became more important than experiments and observations. It is true that Watson and Crick determined the double helix structure of DNA without having carried out a single experiment on the molecule. It is also true that Delbrück had introduced a new style of work into the phage group in which discussion of experiments and articles was brought to the fore.[16] Bacterial genetics involved experiments that were difficult to think up but simple to carry out.

It should, however, be borne in mind that Watson and Crick were exceptions. Experiments in both crystallography (determining protein structure) and biochemistry (studying protein synthesis), which were essential components of molecular biology, required both time and care. In the 1960s, some specialists in bacterial genetics would abandon the subject in order to carry out molecular studies of higher organisms. They had to leave their offices, where they had spent most of their time working out subtle genetic crosses, in order to spend many hours in cold rooms trying to isolate minute quantities of proteins or RNA. A close study of Watson and Crick's approach, however, reveals that their research, like that of their colleagues, was more akin to tinkering than to the work of an engineer or an architect.[17] A builder chooses elements in which he or she has confidence and, on that basis, constructs a stable building. Researchers, by contrast, have only rotten planks available, with a non-negligible probability that they will give way. From these rotten planks they choose a few that they will use to build a new edifice. In most cases, the building collapses, but occasionally it holds. The rotten planks then become more and more solid as building progresses.

This is what makes a historical analysis so difficult. There is a clear tendency, encouraged by scientists themselves, to justify the choice of planks chosen to build the "edifice" (the right theory), to suggest that these planks were clearly the best from the outset.[18] This, of course, is rarely the case. The reason certain planks are chosen and subsequently become more solid is the product of a combination of chance and scientific flair. Watson and Crick were not certain of the structure of bases, the number of polynucleotide chains, the interpretation of the diffraction patterns, or even of the role of DNA in heredity. And yet, despite all that, they proposed a structure that was so elegant that it convinced everyone who studied it. The geneticist and historian Gunther Stent has suggested that Watson and Crick's discovery was akin to a work of art, and he has used the example to point out similarities between scientific and artistic activity.[19]

These two activities are often presented as opposites. In a work of art, the artist is completely free—he or she invents something that has never existed. The scientist, however, is not free—he or she merely reveals concealed truths that, up until now, have been hidden from human eyes. The work of art is unique, inextricably attached to the identity of its creator. Scientific work is the work of a community, and scientific results are often anonymous because the name of their author is of no importance.

Stent is particularly convincing when he shows that the artist is no more free than the scientist in his or her artistic choices: though it is true that all artistic creation is unique, nevertheless artistic schools do exist, within which all works bear a certain resemblance. By contrast, taking the example of the double helix, Stent shows the extent to which this was a unique discovery belonging to Watson and Crick. If they had not existed, the double helix structure would certainly have been discovered, but less elegantly and more gradually. Yet it can also be argued, as indeed Crick has done, that the importance of the discovery came above all from the object discovered—the double helix[20]—and its remarkably simple structure.

Some models of the history of science that are currently dominant equate research with other human activities. They argue that scientific theories are a reflection not of reality but of many human "constructions," having the same status as literary or artistic works. This "constructivist" vision of science may be seductive when applied to the highly abstract models of particle physics, but it is difficult to see how the double helix could be one of many possible human constructions rather than a reflection of reality.[21]

The story of the discovery of the double helix structure of DNA would not be complete without a description of the tragic fate of Rosalind Franklin.[22] Franklin, who died from cancer a few years after the discovery of the double helix, got short shrift from Watson in his autobiographical account. More recently, she has been presented as a female researcher whose problems were typical of those encountered by women in the male-dominated research milieu. Franklin, according to this analysis, was deprived of *her* discovery of the double helix.

This is wrong. The life of a woman scientist is certainly difficult, but, as we have seen, Rosalind Franklin's problems with Maurice Wilkins had a more trivial explanation. Franklin's crystallographic data were certainly very important to the discovery of the double helix structure of DNA,[23] but it is also clear that Franklin did not share Watson and Crick's awareness of the importance of the structure of DNA, nor of the need to explain the auto-replicative ability of genes on the basis of this structure.

Nevertheless, in the mythology that has developed around the discovery of the DNA double helix, Rosalind Franklin has taken on the role of a martyr. This need felt by scientists to ascribe a mythical origin to new disciplines has been studied by historians of science.[24] In the case of molecular biology, Pnina Abir-Am has shown how such a mythical vision can distort the historical account.[25] Such a distortion is not particularly dangerous when it "rectifies" scientists' personal failings in order to turn them into heroes. It is, however, much more problematic when it twists historical reality and renders the history of a discipline incomprehensible. In the case of molecular biology, this mythical vision has a tendency to concentrate the entire history of the discipline into a single discovery. Previous results were merely steps on "the path to the double helix,"[26] from which flowed all subsequent results, including the discovery of messenger RNA and the genetic code. This presentation of events is wrong: the discovery of the double helix was both a motor and a brake on the later developments of molecular biology.

The discovery of the double helix had an enormous impact both on biologists and on the general public. Paul Doty, a chemist specializing in DNA, has explained that, at the end of the 1950s, he saw a button with the letters "DNA" on it. When he asked the vendor what it meant, he was told that it referred, of course, to the gene and that he had better get wise![27] It was not Avery's experiment, nor even Hershey and Chase's, but the discovery of the double helix that convinced the biological community that genes were composed of DNA, and that it was thus the basis of heredity. Similarly, it was through the double helix that the genetic role of DNA entered the textbooks.[28]

The discovery's highly favorable reception should not hide the fact that it was extremely fragile: although the notion of base complementarity was widely accepted, the idea that DNA was a helix was more difficult to believe—the helix would have to unwind during replication, and this "unwinding" posed a major biochemical (and topological) problem that Watson and Crick had barely touched on.

Other models of DNA were also proposed, in which the two polynucleotide chains were parallel, and indeed some crystallographic results appeared to support these heterodox models. It was only in the 1980s, twenty-five years after Watson and Crick's discovery, that the double helix structure was unambiguously demonstrated.

The discovery of the double helix played an extremely ambiguous role in the subsequent development of molecular biology. Although it showed that the series of bases was fundamental for the genetic function of DNA and its replication, it also suggested that knowledge of the structure of molecules alone could lead to an understanding of their function. As will be seen (Chapter 12), RNA also plays an essential role in protein synthesis. Many researchers, including Watson, turned to the structural study of this molecule.[29] The idea of a genetic code, that only the information carried by nucleic acid molecules was important and not their structure, was still far from the minds of most scientists.

Watson and Crick thought that, during DNA replication, the two strands separated and that each would gather bases from its nucleotide-rich surroundings in order to synthesize a complementary strand. Such a model of replication is called "semiconservative" because half the DNA is totally conserved, whereas the other half is synthesized *de novo.*

To clarify matters, but also because they thought that Watson and Crick's model was too simple, Delbrück and Stent outlined the other possible models of replication[30]—"dispersive" replication (in which nothing was conserved, the parent DNA molecules were degraded, and the new DNA molecules were synthesized), and "conservative" replication, in which the two original strands of DNA remained intact and the two strands of the daughter molecule were synthesized *ex nihilo,* using available nucleotides.

In 1941—long before the exact chemical nature of the gene was known—J. B. S. Haldane, a famous enzymologist and geneticist and one of the founders of the neo-Darwinian evolutionary synthesis, had suggested that gene replication could be studied through the use of heavy isotopes such as nitrogen 15 (^{15}N). The idea was that by suddenly changing the composition of the milieu, scientists could distinguish old and new copies of genes.[31] As seen in Chapter 9, radioactive isotopes had already replaced heavy isotopes in the arsenal of biochemists and molecular biologists. The initial experiments to test the semiconservative model of DNA replication were thus carried out with radioactive isotopes—without much success.

In 1954 Matthew Meselson (one of Pauling's ex-students) and Franklin Stahl decided to return to the idea of separating newly synthesized molecules of DNA according to density. After a number of trials, they decided to label DNA with ^{15}N. How could DNA molecules of slightly different densities be separated? On the one hand the medium would have to have a density very close to that of DNA (this could be obtained by dissolving a heavy salt in water—Meselson and Stahl used cesium chloride), on the other hand, a physical method would have to be used that could create a density gradient—ultracentrifugation.

In 1957, Meselson and Stahl finally carried out their experiment,[32] prior to which they allowed bacteria to reproduce for several generations in a medium containing ammonium chloride labeled with ^{15}N. At the beginning of the experiment, the bacteria were diluted in a normal medium containing nitrogen 14 (^{14}N). Meselson and Stahl took samples of the bacteria during the course of the experiment, extracted their DNA, then mixed it with cesium chloride. After twenty hours of centrifugation at 45,000 rpm, the position of the bacterial DNA in the centrifuge cell was noted.

At the beginning of the experiment, the very dense DNA formed a band at the bottom of the cell. After a while another, less dense band appeared. After one cycle of bacterial reproduction, this band was the only one that was visible. It corresponded to the density of a molecule of DNA formed by a heavy chain labeled with nitrogen 15 and a light chain. This result was exactly that predicted by the semiconservative replication model.

Meselson and Stahl confirmed that the medium-density DNA band detected after a single generation was in fact composed of one light and one heavy strand. They heat-denatured the DNA molecule and obtained two bands of slightly different molecular weight, one with a density corresponding to that of a heavy DNA, the other to the density of a light DNA.

The results were clear, "clean," and in perfect agreement with the predictions of the semiconservative model.[33] Furthermore, the value of the result was reinforced by the elegance of the method—the "density gradient"—which was subsequently widely used in molecular biology for separating DNA molecules.

Meselson and Stahl's result was particularly striking because it had been obtained on *Escherichia coli* and not on bacteriophage, thus showing that the semiconservative model of DNA replication was also valid for whole chromosomes.

CHAPTER 12

Deciphering the Genetic Code

The path that led from the discovery of the double helix structure of DNA to the deciphering of the genetic code was so convoluted that a strictly chronological presentation would be incomprehensible. To resolve the problem of the role of genes in protein synthesis, two approaches were possible. Genetically inspired researchers could try to deduce the role of DNA in protein synthesis on the basis of the structure of the DNA molecule and the action of genes in the cell. The other, more biochemical line of attack was to try to create *in vitro* systems that could synthesize proteins. The second approach was to lead to the deciphering of the genetic code.

A number of books, including Watson's and Crick's autobiographical accounts, support the view that the revelation of the structure of DNA almost naturally gave rise to the idea of a genetic code. According to this hypothesis, the succession of nucleotides (bases) in DNA "code" for different amino acids, which in turn are chained together to form proteins. In their second article, dated May 30, 1953, Watson and Crick did indeed write that "the precise sequence of the bases is the code that carries the genetical information."[1] But this formulation merely reflected the fact that the ideas of "code" and "information" were already widespread among molecular biologists—it did not imply that Watson and Crick had a clear idea of the existence of a genetic code. Indeed, they were surprised to receive, after the publication of their articles in *Nature,* a letter from George Gamow, a Russian-born physicist working in the United States, which contained a concrete proposal for a genetic code: a direct correspondence between bases and amino acids.[2] Gamow's idea was that the DNA double helix contained twenty spaces, the form of which depended on the nature of the surrounding nucleotides. These spaces could be "containers" for amino acids. For a sequence of bases to correspond to a protein, covalent bonds would simply have to link the different amino acids.[3]

Gamow's letter surprised Watson and Crick. At the time, knowledge of protein structure was far too vague for biologists to have any clear idea of a strict genetic determinism of the sequence of amino acids. Gamow had been able to accept such a simple hypothesis only because of his ignorance of the field.

Watson and Crick's first reaction to the letter was to try to show that Gamow was wrong. The weak point of his model was his choice of twenty amino acids—more than one hundred different amino acids are found in nature. Some of these are present in all proteins, others are relatively rare and are produced by the modification of other amino acids after protein synthesis. Gamow's division of amino acids into fundamental and derived compounds was thus not entirely felicitous. Crick, using his long experience in the study of proteins, proposed another list of twenty fundamental amino acids that, remarkably, turned out to be right.

Nevertheless, Gamow's proposal was extremely fruitful. Crick adopted the idea of a genetic code—of a correspondence between nucleotides and amino acids. Gamow's suggestion implied that the code might be deciphered directly, without any experiments, without having to study all the intermediate biochemical steps leading from DNA to proteins.

In Gamow's code, each group of three bases, called a triplet (or codon), coded for a different amino acid, but triplets coding for successive amino acids overlapped. This type of code limited the number of possible amino acid sequences. The rare protein sequences that were known at the time led Watson and Crick to reject this code. Gamow and Martynas Ycas then devised a new code, the "combination code," in which the order of bases was unimportant; what counted was their combination. This code allowed for twenty ways of combining four objects—the four bases taken in groups of three. But it was difficult to imagine how the cellular machinery could recognize a combination of bases.

The idea of an overlapping code was thus rejected. However, the problem with a "simple" code was that two letters provided only sixteen possibilities (fewer than the likely number of amino acids), whereas a code with three letters gave sixty-four possibilities—much greater than the number of amino acids. Furthermore, a non-overlapping code raised the problem of the "reading frame." How could the cell know where one triplet ended and the next one began?

In 1957 Francis Crick and Leslie Orgel proposed a solution to these problems.[4] If each nucleotide triplet was read in the right "reading frame," it would code for an amino acid, but if the reading frame were shifted one base, all the triplets would lose their meaning. This meant that there were only twenty meaningful triplets. This perfect relationship between the number of codons and the number of amino acids appeared too good not to be true.

A number of other kinds of codes were proposed, all of which sought to reduce the number of "significant" codons to twenty or to find a simple method of determining the reading frame. When the genetic code was finally deciphered in 1961, all these attempts were shown to have been pointless.[5]

Most of the proposed codes, and in particular that of Crick and Orgel, were based on the assumption that if proteins had a given amino acid composition, nucleic acids had to have a given nucleotide composition. The more that was known of the DNA composition of different organisms, the more it appeared that the amino acid composition of the same protein in different organisms was more or less identical, whereas the nucleotide composition of these different organisms showed substantial variations. In a lecture given in 1959 at the Brookhaven National Laboratory near New York, Crick listed all the difficulties encountered by the code hypothesis, which, at the time, he was ready to abandon.[6]

Although the idea of a genetic code—and especially the theoretical approaches that were adopted in order to characterize that code—had led to a dead-end, the years following the discovery of the double helix reinforced the idea that genes closely controlled the nature of the amino acids that form proteins.

The decisive steps forward in the characterization of proteins that had taken place in the first half of the century have already been described. The concept of the macromolecule had decisively vanquished colloid theory. Following Pauling's work, researchers generally accepted that proteins were formed by making linear chains of amino acids linked by peptide bonds. The protein chain would then fold into a precise conformation that was stabilized by the formation of weak bonds (hydrogen bonds).

This quick overview is not meant to suggest that, in 1950, scientists held today's view that *whatever* its amino acid sequence, a polypeptide chain will *spontaneously* fold into a stable conformation. All biochemists at the time felt that there must be "rules" that would simplify the problem of protein synthesis. Some biochemists felt that these rules must exist at the level of the chaining of amino acids. In 1937, Max Bergmann and Carl Niemann of the Rockefeller Institute had proposed that each protein was formed by $2n \times 3m$ amino acids, each being present at precise positions in $2n' \times 3m'$ copies—n' and m', like n and m, being whole numbers.[7] This mysterious arithmetic was the result of the action of proteinases (what we would call proteases), which were thought to be responsible for protein biosynthesis. Not all biochemists had such a precise model of protein structure. Many of them, however, agreed that there must be rules for assembling amino acids.

In the absence of any discernible regularity in the chaining of amino acids, scientists hoped to find general principles that would govern the three-dimensional structure or folding of proteins. In this context, Linus Pauling's 1950 discovery of the secondary structures of polypeptides, and in particular of the α-helix (see Chapter 11), was especially important. Pauling thought he had shown not so much several possible models of how proteins folded, as the fundamental structures on the basis of which folding took place.

This certainty that the formation of proteins could be understood simply had its roots in the biochemists' view of the role of genes in protein synthesis. Contemporary theories of protein synthesis were relatively vague: several models existed, all of which were, at least in part, mutually incompatible. Nonetheless, they all gave an essential but limited role to genes. This implied that protein synthesis should be a relatively simple biochemical process.

The hopes of finding any regularity in the chaining of amino acids evaporated, however, when the first protein sequences were established. Only a few years earlier, Synge, Martin, and their colleagues in the Leeds Wool Industries Research Association[8] had developed chromatographic techniques for separating amino acids and peptides, first on silica, then on paper (see Chapter 9). The British biochemist Frederick Sanger had subsequently developed methods for fragmenting polypeptide chains under the action first of acids, then of proteases (and in particular trypsin), followed by their separation and sequencing.[9] The first complete protein sequence was that of insulin.[10] This sequence showed that amino acids had precise positions in the polypeptide chain, but did not reveal any regularity in their sequence.

The virtually simultaneous demonstration that genes control both the nature and the position of the amino acids in polypeptide chains was made in two steps, separated by more than seven years. The first step was made by Linus Pauling. In a 1949 article published in *Science,* Pauling had shown that sickle cell anemia—so called because patients suffering from the disease have sickle-shaped red blood cells—was linked to an abnormal structure of hemoglobin, the protein that transports oxygen in the blood.[11] Pauling showed that the hemoglobin of sickle cell anemia patients had an electric charge different from that of normal hemoglobin. Furthermore, at low oxygen pressures it was less soluble. Under these conditions, "sickle cell" hemoglobin precipitated in the form of long needles, deforming the red blood cells and giving them their characteristic sickle shape. These red blood cells could not fit properly in the patient's capillaries and thus tended to block the oxygenation of the tissues. More fragile than normal red blood cells, they produced anemia in the patient. Shortly before Pauling's discovery, James Neel, a specialist in human genetics at the University of Michigan, had shown that sickle cell anemia was a genetic disorder displaying Mendelian inheritance.[12]

Pauling's article was extremely important for the history of modern medicine: by showing how a molecular disorder can explain the symptoms of an illness, it founded molecular medicine. Pauling and his colleagues wanted to go on to characterize the modification(s) of hemoglobin that resulted in the observed new properties, but preliminary studies showed that normal and mutant hemoglobin had the same amino acid composition, and so they abandoned their attempts.

The second step demonstrating that genes control both the nature and the position of the amino acids in polypeptide chains was made by the British scientist Vernon Ingram, at Crick's insistence. Following a visit to the Rockefeller Institute, Ingram began work at the Cavendish Laboratory in Cambridge, using Sanger's protein sequencing technology. In 1956 he showed that the mutation responsible for sickle cell anemia was linked to a change in a single peptide and to the replacement of a single amino acid in this peptide.[13]

Ingram's discovery had an important impact on biochemists: it showed that genes intervened directly in protein structure and influenced such "unimportant details" as the nature and position of a single amino acid.

Up until this point the biochemical problems associated with protein synthesis have only been hinted at. The problem of which mechanisms are involved in protein synthesis was initially considered to be the opposite problem to degradation. Because enzymes—proteases—were able to cleave proteins at specific sites, identical or analogous enzymes were thought to carry out the opposite operation. Furthermore, in addition to direct synthesis reactions involving polypeptide chains, there might be reactions involving the exchange of amino acids between peptides, which were energetically more favorable. Schoenheimer's studies using nitrogen 15 had shown that proteins were in an unstable metabolic state, and that amino acids were continually added to or subtracted from them (see

Chapter 9). These results agreed with the proposed role of proteases in protein synthesis.

Between 1940 and 1955 the multi-enzyme model of protein synthesis was widely accepted.[14] It was, however, progressively undermined:

- It presented a major theoretical difficulty: if each protein, each enzyme, was synthesized by a multi-enzyme complex, how were the enzymes in the multi-enzyme complex synthesized? If they were themselves synthesized by multi-enzyme complexes, it was difficult to see how the whole process could ever begin.[15] Furthermore, the one protein–one multi-enzyme complex did not agree with Beadle and Tatum's results and the one gene–one enzyme hypothesis (Chapter 2).
- The multi-enzyme model of protein synthesis left no place for nucleic acids. Even if the precise role of nucleic acids remained unclear, they nevertheless appeared to intervene at two levels. From the late 1940s molecular biologists knew that, in one way or another, genes controlled protein synthesis and that they were formed, at least in part, of nucleic acids. By contrast, researchers on protein synthesis in eukaryotic organisms (rather than in bacteria), in which chromosomes are isolated from the rest of the cell by the nuclear membrane, had shown that protein synthesis takes place in the cytoplasm and not in the nucleus, where the genes are.[16] At the beginning of the 1940s, research by the Swede Torbjörn Caspersson and by the Belgian Jean Brachet had shown a correlation between the level of protein synthesis and the quantity of RNA in the cytoplasm[17] (see Chapter 13). Once again, nucleic acids were implicated in protein synthesis.

In 1952 Alexander Dounce of the University of Rochester, well aware of these problems of the multi-enzyme model, proposed the radically different template model.[18] According to this model, nucleic acids were the scaffolding on which the amino acids were assembled in order to form proteins. "P1" enzymes made the link between nucleotides and amino acids, whereas nonspecific "P2" enzymes formed the peptide bonds between the different amino acids that had been brought together. Despite its appeal, this model left a fundamental question unanswered: what was the template? If it was DNA, why were proteins synthesized in the cytoplasm, whereas genes—DNA—were found in the nucleus? And if it was RNA, how could the role of genes in protein synthesis be explained?

To answer these questions, Dounce suggested in 1953 that DNA might itself serve as the template for the synthesis of RNA, and that RNA was the template for protein synthesis.[19] This was a remarkable anticipation of what was to become the central dogma of molecular biology (see Chapter 13). But strictly speaking the theory was not new—it had been proposed in 1947 by André Boivin and Roger Vendrely.[20] The nature of the RNAs involved in protein synthesis remained unclear, and the originality of Dounce's model was not widely recognized. For Dounce, P1 enzymes alone were responsible for making the link between an amino acid and a given nucleotide in a polynucleotide sequence. He was the first person to consider the relation between proteins and nucleic acids to be indirect and thus nonstereospecific (see Chapter 1) and structurally arbitrary. But this was not how the template model was understood. Scientists interpreted it in the light of biochemical tradition, according to which the formation of biological macromolecules was the result of their stereospecific interaction with other molecules. The archetype of such models was Pauling's 1940 explanation of antibody formation.[21] A description of this model will make it clearer how much molecular biology differed from previous approaches.

When a foreign molecule—an "antigen"—enters a higher organism, the organism reacts by synthesizing proteins called antibodies, *in virtually all cases* and whatever the chemical nature of the antigen.[22] The antibodies fix on the antigen, thus forming a complex that can be eliminated.

Chapter 9 showed how the understanding of antibodies had rapidly developed through the use of the new physical techniques of electrophoresis and ultracentrifugation. These methods had shown that antibody molecules were γ-globulins and all had more or less the same amino acid composition. Pauling, using detailed studies on the effects of chemical modifications of antigens carried out by the Austrian-born immunologist Karl Landsteiner[23] at the Rockefeller Institute, was able to describe precisely the chemical nature of the bonds that formed between the antibody and the antigen (see Chapter 1).[24] These studies explained how the interaction between the antibody and the antigen took place, but said nothing about the origin of antibodies. How could the organism produce antibodies with molecular structures that complemented literally any foreign molecule?

The simplest solution had been suggested as early as 1930 by Stuart Mudd, Jerome Alexander, F. Breinl, and Felix Haurowitz:[25] antibodies were able to interact with antigens because the antigen guided the formation of the antibody. The antigen had an "instructive" role.

In 1940 Pauling adopted this model and gave it the chemical precision it lacked. For Pauling, all antibody molecules had identical amino acid chains. They were, however, very unusual proteins: the central part was highly structured, but the two extremities appeared not to have a definite form and "hesitated" between various conformations. This effect might be due to the presence of a particular amino acid, proline, in these regions. Newly synthesized antibody molecules do not immediately take on their final conformation—when in the presence of an antigen each extremity folds around the foreign molecule and adopts the definitive conformation—complementary to that of the antigen. The antigen then escapes from the antibody molecule, thus resulting in the production of antibodies ready to interact with any new molecule of the antigen that they might encounter.

This theory was thoroughly "scientific" in that it could easily be put to the test of refutation: it led to a certain number of predictions that were open to simple experimental testing:[26]

- Because an antibody adopts its final conformation only in the presence of an antigen, the synthesis of specific antibodies should stop if the antigen is eliminated from the organism. But it was already known that immunity could persist for many years, while, as Schoenheimer had shown, proteins—which included globulins—showed a rapid turnover. This suggested that the antigen was stored somewhere in the organism. By using a radioactively labeled antigen, scientists could track its fate following injection. Experiments showed that though most of the antigen rapidly disappeared, a small quantity remained.[27]

- Pauling had shown that antibodies were bivalent, with two antigen-recognition sites. If an animal was immunized with a complex antigen containing two different chemical groups, a and b, one would expect to obtain antibodies with a dual specificity, directed against both a and b. Pauling cited a number of results that partly agreed with this prediction.
- Above all, Pauling's model suggested a way of making antibodies *in vitro.*

These experiments, which Pauling announced in 1940, were carried out with his colleagues and published in 1942;[28] they succeeded beyond his wildest dreams:

- Pauling first used a dye as an antigen. Various immunoglobulins from a nonimmunized animal were added to the dye (he denatured the immunoglobulins by passing them through an alkaline medium). The globulin-dye mixture was then slowly brought to a neutral pH, forming a precipitate of the antigen and immunoglobulin.
- Pauling then used a milder procedure for denaturing the immunoglobulins—heating to 10°C below the antibody denaturation temperature. This temperature should have speeded up the conformational changes in the immunoglobulin molecule. More than 40 percent of the globulins developed an antibody function directed against the antigen.
- Still using this less vigorous denaturing procedure, but this time with a polysaccharide extracted from a pneumococcus as an antigen, Pauling obtained virtually identical results, although there were problems with the experiment owing to difficulties in eliminating the antigen from the precipitate without denaturing the antibodies.

Of course, the antibodies obtained in these three experiments precipitated the antigens in conditions that were slightly different from those required by natural antibodies extracted from an animal. But this difference was small and the restriction appeared relatively minor compared with the overall outcome, which completely agreed with Pauling's model.

Pauling was sufficiently encouraged by these results to consider producing specific antibodies on an "industrial" scale. The Rockefeller Foundation clearly understood the importance of these results, and provided Pauling with substantial financial support.

Nevertheless, these initial successes were not immediately replicated. Felix Haurowitz, who worked in Istanbul, criticized one of Pauling's experiments by showing that precipitation of globulin molecules and dyes coupled with ovalbumin could occur spontaneously, as a result of the attraction of opposite electrical charges.[29] Haurowitz's criticism, however, was only valid for one of Pauling's many experiments. The rest of his data, which showed the *in vitro* formation of antibodies against a single dye molecule or against pneumococcal polysaccharides, were not in question. Despite his criticisms, Haurowitz in fact agreed with Pauling's theory of antibody formation.

Pauling's first results could not be reproduced.[30] As is often the case in science, these negative data were not published. More curiously, Pauling's model is cited in the scientific literature as if it remained purely theoretical.[31] The experiments, which nevertheless provided a striking confirmation of the model, are never mentioned, even by their authors.[32] Pauling's 1940 model of antibody synthesis was later replaced by the model of clonal selection developed by Burnet in 1957 (see Chapter 17). Nevertheless, the 1940 model was of fundamental importance for biochemistry.

Pauling believed that his model was valid only for antibodies, which he thought were the only proteins to show conformational flexibility, the form of other proteins being strictly determined by their amino acid sequence. But this model was implicitly or explicitly extended to all proteins and cellular enzymes. For many biochemists, the substrate of an enzyme—the chemical compound that the enzyme transforms—intervened in the synthesis of the enzyme. Some imagined that enzymes took on their definitive conformation by folding around a substrate molecule.[33]

This theory was particularly well suited to the adaptive molecules studied by Jacques Monod and Sol Spiegelman (see Chapter 14). These enzymes were produced only when organisms were in contact with the relevant substrates—it was thus logical to imagine that the substrate played a role in synthesis. In 1947 Monod proposed a model that showed how the substrate converted an inactive protein into an active enzyme, almost certainly by conformational change.[34] For Monod, this "induction" was a general phenomenon, closely linked to protein synthesis.[35] All enzymes could be induced; those enzymes that appeared to have a constitutional activity were, in fact, synthesized under the action of endogenous inducers that were always present in the cell. The model proposed for inducible enzymes was thus generally valid.

The template model and Pauling's model might appear different, if not incompatible: the first suggests that protein structure is the product of a kind of copy, the second that this structure is obtained when the antibody molds around another, "negative" molecule. But these models were never formally confronted. The term "template" was sufficiently vague to mean both "pattern," suggesting a process in which an identical copy was produced, and "mold," implying a process in which a complementary copy was produced. For most biochemists and geneticists, the two models were part of the same biological vision, according to which form played a fundamental role. As seen earlier, this idea of form, of specificity, taken from enzymology and immunochemistry, played a major and complex role in the birth and development of molecular biology.[36]

The study of these two models is particularly difficult because, although they were omnipresent in the thinking of biochemists from the 1940s to the 1960s, they were almost never explicitly formulated. It is as though scientists thought that if their models were set down on paper, their fragility would be exposed. Sometimes, however, in interviews or popular articles, scientists expressed themselves more freely: for example, in a 1949 interview with *Scientific American,* George Beadle described the development of molecular biology at Caltech in the following terms:

> We are seeking to uncover the principles that govern fundamental processes of life . . . the investigations tend to show that the molecular form known as protein is the key structure . . . The genes, we believe, exercise an overruling control on all these activities. They do this, we think, by serving as the master patterns for the many proteins which function in the processes of life. Thus, there is probably a gene which serves as the template for the body's manufacture of insulin, another which provides the mold for pepsin, and so for albumin, fibrinogen, the polypeptide chain that forms antibodies, and all the rest.[37]

A final example will give an idea of the close links that existed between the concepts and tools of immunology and those of other biological disciplines—biochemistry and even genetics.

In 1944 the geneticist Alfred Sturtevant,[38] who had worked for many years with Thomas Morgan, linked two sets of data. One was the work of immunologists, who had shown that some antigens, such as blood groups, were the direct product of gene action;[39] the other was the work of Landsteiner followed by Pauling, who had shown that antibodies and antigens had complementary structures. Sturtevant deduced that antibodies directed against an antigen would probably interact with the gene responsible for the synthesis of the antigen and that it would thus be possible to induce specific mutations. This is what Sterling Emerson had just succeeded in doing in *Neurospora*.[40] The data were extremely important. When they were not confirmed, however, they were forgotten, together with the models that they supported.

Fortunately, biology is a science and all models or hypotheses fade away in the glare of contradictory experimental results. In the case of protein synthesis, these results were the product of the patient work of biochemists who tried to reproduce this synthesis *in vitro*.

But the active role of biochemists in what was to be one of the most remarkable discoveries of molecular biology—the deciphering of the genetic code—is not widely known.[41] This fact, and the consequent resentment felt by biochemists, can be traced to the first descriptions of the birth of molecular biology, particularly those written by the contributors to a 1966 festschrift for Max Delbrück, in which molecular biology was presented as the fruit of the collaboration of the phage group geneticists and the Cambridge crystallographers and structural chemists.[42] The role played by Watson and Crick in the discovery of the double helix structure of DNA was taken as the symbol of this collaboration.

Watson's book *The Double Helix* reinforced this bipolar vision of the history of molecular biology. Watson tended to ignore the role of biochemists—for example, the importance of the crystallization of the TMV or, as discussed in this chapter, studies of protein synthesis.

After the war, using new experimental possibilities provided by the production of radioactive compounds, several groups directly studied the mechanisms of protein synthesis. One of the most active laboratories was led by Paul Zamecnik at Harvard.[43] The initial aim of Zamecnik's group was to compare protein synthesis in normal and cancerous cells. The first experiments were carried out on slices of rat liver, some of which had been treated with a cancer-inducing agent that provoked the formation of hepatomas. This required the development of some very precise techniques—obtaining radioactive amino acids, measuring the radioactivity incorporated into proteins—and led to some very disappointing results: there were differences between normal and cancerous tissues, but they were merely quantitative and not qualitative and thus did not alter the nature of the proteins that were synthesized.

These results led to a reorientation of the research carried out at the Zamecnik lab: the group now sought to open the "black box" of protein synthesis. This change was stimulated by a novel observation made by Zamecnik's group: protein synthesis appeared to depend on the energetic state of the cell. This did not agree with the multi-enzyme model of protein synthesis that was dominant at the time. Furthermore, in 1950, Henry Borsook showed that protein synthesis took place in a particular cell structure, the microsome. Borsook's observation allowed the comparison of *in vitro* and *in vivo* data and therefore provided a criterion for testing the validity of the former.

As early as 1951, Zamecnik succeeded in obtaining a coarse fractionation of a homogenate of liver cells in which the incorporation of radioactive amino acids into proteins could be observed *in vitro.* In the years that followed, Zamecnik's group developed protocols that made it possible to distinguish between radioactive amino acids involved in "real" protein synthesis and any "artifactual" incorporation. Following a series of increasingly powerful fractionations, the system became sensitive enough to be useful in studying the key stages of protein synthesis. In 1954 Zamecnik developed an active *in vitro* system that contained only amino acids, a "donor" molecule (ATP) to provide energy, microsomes, and a supernatant from a high-speed ultracentrifugation of a raw cellular extract.

This system showed that ATP was necessary for protein synthesis, which was a severe blow for the multi-enzyme models, in which there was no place for ATP. For biochemists trained in the study of metabolism, the importance of ATP implied that there must be an amino acid form that was activated through a reaction with ATP. The Belgian biochemist H. Chantrenne proposed a structure for this activated form, which was rapidly confirmed by Mahlon Hoagland in Zamecnik's lab.[44] But this was merely the first step: Hoagland showed that the amino acid was subsequently loaded onto a small RNA, which he called soluble RNA. The same enzymes were responsible for the fixation of the amino acid to ATP, then to soluble RNA. There were as many enzymes as amino acids.

These biochemical discoveries validated the hypothesis that Francis Crick had put forward a year earlier, in 1954, according to which there must be small RNA molecules, called "adaptor" molecules, able to fix amino acids and to interact with the nucleic acid matrix.[45] This hypothesis had been inspired by Gamow's model of the genetic code. One of the weaknesses of this model was that it could not really describe in "chemical" terms how a nucleic acid could form a "container" for the lateral chains of amino acids. On the other hand, the discovery of the DNA double helix had shown that one nucleic acid could combine with another nucleic acid by forming a group of hydrogen bonds. Crick hypothesized that there were twenty adaptor RNAs, one per amino acid, and twenty enzymes able to bind specifically one type of amino acid to an adaptor RNA. The adaptor RNA would then bind to the matrix of nucleic acids.

The soluble RNA molecule (several dozen nucleotides) was much bigger than Crick had imagined. Nevertheless, this difference between the expected and the observed values did not produce any serious problems.

The convergence between the most theoretical approach to deciphering the genetic code and the biochemists' "ground-level" approach confirmed the experimental value of *in vitro* models of protein synthesis. Shortly afterward, Zamecnik developed a system of *in vitro* protein synthesis using bacterial extracts[46] that was subsequently improved by Alfred Tissières.[47] Everything was in place for the deciphering of the genetic code.

On Monday, May 22, 1961, at 3:30 P.M., Johann Heinrich Matthaei, a German biologist working in the United States, took a test tube and mixed a ground-up extract of bacteria that had been centrifuged, a fraction containing small molecules of soluble RNA, the twenty amino acids that form proteins (sixteen of which were radioactively labeled), ATP as an energy source, salts, and a buffer that kept the pH of the mixture constant.[48] He then added a few micrograms of an artificial molecule of ribonucleic acid that had been synthesized *in vitro,* consisting of the simple repetition of one type of nucleotide (UMP). After incubating the mixture for one hour at 35°C, Matthaei precipitated the proteins from the mixture using trichloroacetic acid, washed the precipitate, and placed it in a radioactivity counter. The result was clear: in the presence of the "poly-U" nucleic acid, but not in its absence, amino acids were incorporated into a material that could be precipitated in an acid medium, that is, into proteins. The poly-U had led to the synthesis of a protein, entirely *in vitro.*

For the rest of the week, Matthaei worked day and night to determine which amino acid(s) was or were incorporated into proteins in the presence of poly U. On Saturday, May 27, at 6:00 in the morning, he finally had the answer: the poly U coded for a monotonous protein consisting of a chain of a single amino acid—phenylalanine. In less than a week, Matthaei had identified the first "word" of the genetic code.

Marshall Nirenberg, who directed the small group at the National Institute of Arthritic and Metabolic Diseases at Washington in which Matthaei worked, came back from a four-week visit to Berkeley. Because Matthaei had to stop work for two weeks to attend a course on bacterial genetics, Nirenberg finished characterizing the product of the experiment. The two researchers then quickly wrote two articles, one describing the modifications they had made to the system of *in vitro* protein synthesis,[49] the other presenting the results obtained with different RNAs, including the poly-U molecule.[50] These two articles, sent on August 3, 1961, to the *Proceedings of the National Academy of Sciences,* were published in November of the same year.

Even prior to publication, Nirenberg and Matthaei's results were widely known: they had turned into the main event of the Fifth International

Congress of Biochemistry held in Moscow in August 1961. Neither Matthaei nor Nirenberg was well known, and the institute where they worked was not particularly famous. Furthermore, Nirenberg had not been accepted as a participant at a Cold Spring Harbor colloquium held in June 1961. But international congresses such as the Moscow meeting were more open to scientists than relatively closed meetings such as those at Cold Spring Harbor, and young researchers could present their results. Nirenberg had fifteen minutes to present his data. His talk was scheduled for a small room, and virtually nobody was there to hear it. One of the few people in the room told Crick of the potential importance of Nirenberg and Matthaei's discovery. After discussion with Nirenberg, Crick invited him to present his data again, but in the main conference room. This time the whole conference was electrified by the data. In the weeks that followed, several other groups reproduced Nirenberg and Matthaei's results and extended them to other synthetic RNAs.

The results showed the superiority of the experimental method over theoretical approaches: one of the strongest predictions of Crick and Orgel's 1957 model was precisely that the words of the genetic code formed by a single "letter" (only one kind of nucleotide, for example, U) would not code for anything—they would be "nonsense" words.

Matthaei and Nirenberg's experiment was extremely important. But what was their real contribution? Their study was part of a biochemical tradition aimed at showing that biological phenomena are merely extraordinarily complex physico-chemical phenomena that can be understood when they are reproduced *in vitro.* Zamecnik had been the first to develop cell-free systems of protein synthesis, whereas most of the changes introduced by Nirenberg and Matthaei had already been developed by Tissières. Nirenberg and Matthaei had nevertheless been able to reduce the noise in the system.

But Matthaei and Nirenberg had dared to take the idea of a genetic code to its logical conclusion and to try to determine this code experimentally, without worrying about the precise nature of the RNA involved in protein synthesis (see Chapter 13). They also rejected the idea, deeply rooted in the biochemists' view of the world but rarely openly expressed, that the form of RNA molecules played an essential role in protein synthesis.

There has been a persistent rumor among molecular biologists that Matthaei and Nirenberg's success was in fact pure luck. The poly U RNA had been used in the *in vitro* synthesis medium as a negative control: artificial, formless RNA was assumed to be noncoding. Matthaei and Nirenberg have protested against this view of their work, and a number of historians of science have supported them.[51] Whatever the case, they can be credited with having immediately and correctly interpreted the data obtained with poly U (even if these data were not what they expected!).

The rest of the code was rapidly deciphered. In 1966 all the codons—the base triplets that correspond to the different amino acids, as well as the punctuation signs—had been characterized. The deciphering of the genetic code elicited great enthusiasm and was followed by the major U.S. newspapers virtually day by day. Severo Ochoa's laboratory, already a leader in the study of polynucleotides, threw itself into the race. Other laboratories joined in; data accumulated, although their quality did not match the speed with which they were obtained.[52]

The first polynucleotides used were composed of either a single nucleotide or several nucleotides that, under the action of an enzyme (polynucleotide phosphorylase), were randomly distributed in the polynucleotide chain. On the basis of such polynucleotides, however, a given codon could not be paired with a given amino acid. This problem was overcome by Gobind Khorana, who developed an effective chemical method for synthesizing polynucleotides with a given sequence.[53] Nirenberg also showed that very short polynucleotides (containing only three nucleotides) were sufficient to fix onto the microsomes (now called ribosomes) and to bind the soluble RNA (now called transfer RNA) to them, together with the attached amino acid,[54] thus providing a very simple way of deciphering the code.

Crick and Brenner's elegant approach to the problem of the genetic code,[55] which combined a genetic analysis with protein sequencing, came too late. It merely confirmed that the code was composed of three-letter "words" and that the genetic message was linear.[56] Nevertheless, these genetic techniques were useful in confirming that certain codons were indeed "nonsense" codons, their only function being to interrupt protein synthesis.

When researchers began deciphering the genetic code, the link between DNA, proteins, and RNA, and the role of the different RNAs in protein synthesis, were unknown. Scientists had to discover messenger RNA before they could answer these questions.

CHAPTER 13

The Discovery of Messenger RNA

George Gamow had suggested that different amino acids fit into different "holes" in the DNA molecule. He began his research by looking for a structural relationship between genes (formed of DNA) and proteins. By contrast, biochemical studies of protein synthesis tended to emphasize the role of RNA. The coexistence of these two lines of research shows that the roles of DNA and RNA were not clearly defined. Francis Crick has claimed that in 1953, when he and Watson discovered the DNA double helix, they already had a relatively clear idea of the sequence DNA→ RNA → proteins.[1] Articles and documents from the time show, however, that this was not the case and that right up until 1960, the relation among the three kinds of macromolecules remained extremely vague.

In 1957 Crick gave a lecture to the British Society of Experimental Biology entitled "On Protein Synthesis," which was subsequently considered the key exposition of the "dogmas" of molecular biology (see Chapter 15). In this talk, Crick argued for the first time that protein folding was a spontaneous process and that the final conformation was simply a function of the amino acid sequence.[2] He also proposed the sequence hypothesis, according to which the specificity of a nucleic acid resided only in its sequence of bases—the code that determines the amino acid sequence in the protein. He also stated the "central dogma" of molecular biology: information can go from a nucleic acid to a nucleic acid, and from a nucleic acid to a protein, but not from a protein to a protein, or from a protein to a nucleic acid. Finally, he reintroduced the adaptor hypothesis (see Chapter 12) and set out the "coding problem" and the various theoretical attempts to resolve it.

From a biochemical point of view, Crick emphasized the importance of RNA-rich microsomal structures ("ribosomes") in protein synthesis. But though he devoted several pages to proving that genes—DNA—control the amino acid sequence, and spent some time explaining the role of microsomes in protein synthesis, he devoted only one line to the relation between DNA and RNA: "the synthesis of at least some of the microsomal RNA must be under the control of the DNA of the nucleus" (p. 153).

Before adding natural or synthetic polynucleotides to their cell-free systems, Matthaei and Nirenberg had tried to see if the spontaneous protein synthesis that occurred in these systems required the presence of DNA or of RNA. To each system they had thus added either DNase, an enzyme that could degrade DNA, or RNase, an enzyme that could degrade RNA. *Both enzymes* inhibited protein synthesis. The action of RNase was quicker, perhaps suggesting a passage from DNA to RNA.[3] Matthaei and Nirenberg were very cautious in their conclusions as to the roles of the two macromolecules.

The precise relation between DNA and RNA was thus the last of the relations between the three fundamental biological macromolecules to be understood. It was clarified through the work of the two French scientists François Jacob and Jacques Monod and their discovery of messenger RNA.

The proof that RNA was involved in protein synthesis went back many years and was the result of a set of observations made on very different systems. The first experiments were carried out in the 1940s by Torbjörn Caspersson and Jean Brachet. Caspersson[4] had developed some extremely sensitive spectroscopic methods that enabled him to measure the absorption of ultraviolet light by different components of the cell. Linked to the action of different degrading enzymes and dyes, these studies had shown that chromosomes were composed of nucleoproteins and that the cytoplasm was rich in ribonucleic acids. The level of RNA was proportional to the metabolic activity of the cell—to protein synthesis.

Brachet's approach was different, in terms of both the problem he sought to unravel and the methods he employed.[5] Brachet wanted to understand the role of nucleic acids in embryonic development, using biochemical and histochemical methods. His experiments gave the same result as Caspersson's: the amount of DNA increased in proportion to cell division, whereas the amount of RNA was related to the protein synthesis activity in the cells.

The experiments that most clearly showed that RNA was sufficient for protein synthesis were Brachet's 1955 studies involving anucleation (the removal of the cell nucleus). Protein synthesis continued for several days in anucleate cells, in the absence of DNA.[6] Parallel to these studies, researchers at the Rockefeller Institute had carried out experiments involving cell fractionation by ultracentrifugation. The initial motives behind these experiments were far removed from the problem discussed in this chapter.[7] In 1911, Peyton Rous of the Rockefeller Institute had transferred a cancer from one chicken to another via a cell-free extract he called a virus. The American James Murphy and the Belgian Albert Claude had tried to purify Rous's tumoral agent. Although they succeeded, in a control experiment they also isolated small particles—identical to the virus—from normal, noninfected cells. They hypothesized that Rous's tumoral agent was the result of the autocatalytic transformation of an endogenous component of the cell. They thus adopted John Northrop's theory of bacteriophage replication (see Chapter 4). These endogenous particles were distinct from the mitochondria and were called microsomes.

But in 1943 Claude abandoned the idea that microsomes were related to tumoral agents, arguing that they were instead self-replicative particles that were present in the cytoplasm. He dropped all reference to his previous views and henceforth presented his work as being aimed at understanding the complex inner workings of the cell.

The scientist responsible for the characterization of microsomes was the Rumanian biologist George Palade,[8] who used the electron microscope to study cell cultures and thin sections of tissues, together with biochemical analyses and ultracentrifugation. After extensive comparison of the results from these different techniques, which were often contradictory, Palade was able to distinguish particles—"ribosomes"—from the membranes with which they were associated. Despite this finding, the two terms "ribosome" and "microsome" continued to be used indiscriminately for several years; the RNA that makes up ribosomes was often called "microsomic RNA." Each ribosome was shown to be around 250 angstroms in diameter, and to contain equal amounts of RNA and proteins. Ribosomes were found in all tissues, in quantities that were proportional to the amount of protein synthesis. In 1950 Henry Borsook and his colleagues discovered that protein synthesis takes place in microsomes; in 1955, using a cell-free system, Paul Zamecnik showed that the incorporation of amino acids into proteins takes place on ribosomes.

In addition to these experiments, studies on the tobacco mosaic virus thus confirmed the role of RNA in protein synthesis. In 1956 Heinz Fraenkel-Conrat and Gerhard Schramm (finally) showed that it was the viral RNA that was infectious (see Chapter 6). Fraenkel-Conrat separated RNA from viral proteins; by combining them again, he was able to obtain an infectious virus. Using the same strategy, but combining RNAs and proteins from different viral strains, he had shown that the virus produced by an infected plant was of the same strain as the RNA, not the proteins. RNA clearly controlled the synthesis and nature of viral proteins.[9]

Messenger RNA was discovered in 1960 by François Jacob and Jacques Monod during a study of an "inducible" enzyme, β-galactosidase (see Chapter 14). Monod had characterized two classes of mutations that affected the production of the enzyme: one class prevented the synthesis of an active enzyme, whereas the other led to permanent, "constitutional" synthesis of the enzyme, irrespective of whether or not the inducer was present.

In order to understand how the second class of mutations produced their effects, Jacob and Monod used bacterial sexual reproduction, discovered ten years earlier by Joshua Lederberg. In crosses between male and female bacteria, the male transmits to the female a chromosomal fragment but no other cellular components (see Chapter 5).

An observation made during these experiments was to prove fundamental for the birth of the concept of messenger RNA. These experiments clearly showed that as soon as the β-galactosidase gene penetrated a female bacterium, enzyme synthesis began at a maximum rate.[10] But the male bacterium was thought to transmit no microsomal particles to the female. The result of the experiment was surprising, and contradicted everything that was known about protein synthesis and the role of microsomal particles.

Arthur Pardee, François Jacob, and Jacques Monod carried out a large number of control experiments in order to eliminate any biases in the results. Pardee, who had spent a sabbatical year in France, returned to the United States, where, together with Monica Riley, he conducted an experiment whose result was complementary to that of the experiment he had performed with Jacob and Monod. Pardee and Riley introduced radioactive phosphorus into the bacteria; as the phosphorus decayed, it inactivated the bacterial genes. They showed that as soon as the gene was destroyed, β-galactosidase synthesis stopped. The gene thus closely controlled protein synthesis.[11] Despite this result, Jacob and Monod did not think that the protein was directly produced on the gene. Although this was theoretically possible in bacteria, it was known that in higher organisms genes were confined to the nucleus, whereas proteins were synthesized in the cytoplasm. Jacob and Monod therefore hypothesized that there existed a short-lived intermediary between genes and microsomal particles. This intermediary was probably an RNA, which they first called "X" before naming it "messenger RNA" in the fall of 1960.[12] But microsomal RNA was stable; "X" had to be different. The data from the "PaJaMo" experiments, named after their authors—Arthur Pardee, François Jacob, and Jacques Monod—agreed with results obtained at the same time by François Gros, who was also working in Monod's laboratory, but using a completely different experimental approach. Gros had shown that the addition of a nucleotide analogue, 5-fluorouracil, led to the virtually instantaneous blocking of β-galactosidase synthesis.[13]

Jacob described his data to Crick and Brenner during a visit to Cambridge on Good Friday, 1960.[14] During the discussion, Crick and Brenner suddenly realized that Jacob and Monod's results were identical to those previously obtained by Elliot Volkin and Lazarus Astrachan, which up until then had remained unexplained.[15] Volkin and Astrachan had studied replication of bacteriophage T2 in *E. coli* and had shown that an RNA, comparable to the phage DNA, was synthesized shortly after infection. They deduced that this RNA must be necessary for bacteriophage replication. The results of the Pasteur Institute group shed new light on these data and provided a much more precise interpretation: the RNA that Volkin and Astrachan had observed was a messenger RNA that controlled the synthesis of proteins required for the phage to replicate.

Everything appeared straightforward: the RNA contained in the microsomal particles was not messenger RNA, which controlled protein synthesis. Ribosomes were merely "reading heads" to which messenger RNAs would bind in order to be translated into proteins. This kind of RNA had hitherto gone unnoticed; it was short-lived and of varying sizes, and represented only a small fraction of cellular RNAs. It was also much less common than ribosomal RNA.

The discussions continued into the evening and through the reception organized by Crick. During this gathering Jacob and Brenner developed an experimental strategy for proving the existence of this new type of RNA and for distinguishing it from ribosomal RNA. These experiments were performed in the summer of 1960, in Meselson's laboratory in California.[16] Similar experiments were carried out by François Gros in Watson's laboratory in the United States.[17]

Jacob and Brenner showed that RNA synthesized by bacteriophages that infected a bacterium would bind to the ribosomes that were present in the cell prior to infection. To carry out this experiment, they used the density-gradient technique developed by Meselson a few years earlier (see Chapter 11). The bacteria were placed for several generations in a medium rich in nitrogen-15 and carbon-13. These two heavy isotopes were incorporated into the bacteria and in particular into the ribosomes, making them denser. At the beginning of the experiment, the bacteria were placed in a medium of normal density that contained phosphorus 32 and would label the nucleic acids, and then they were infected with phage. The experiment showed that the newly synthesized labeled RNA molecules bound to the heavy ribosomes. The ribosomes thus merely played a passive role in the synthesis of phage proteins: they were the material support upon which the short-lived RNA molecules were bound in order to be translated into proteins.

François Gros's experiment was "technologically" less sophisticated, but it had the advantage of being carried out on normal, non-infected bacteria. After adding radioactive elements that could be incorporated into the nucleic acids, Gros fractionated the RNA molecules by sucrose gradient centrifugation, which separated molecules according to their size. When the radioactive components were added for only a short time, a new family of RNA molecules was found, different from ribosomal RNA and of varying sizes. But these RNA molecules were still attached to the ribosomes. These rapid-turnover RNA molecules—they were strongly labeled despite the short duration of the experiment—had all the expected properties of messenger RNA.

According to Francis Crick, the discovery of messenger RNA was postmature.[18] But for an unfortunate series of circumstances it would have happened earlier: the fact that the most abundant RNA in cells is ribosomal RNA and that this RNA has a structural function was "bad luck." If ribosomes were made of proteins, and if the only cellular RNA, apart from the small molecules of soluble RNA, had been messenger RNA, it would have been a lot easier to discover!

By the end of the 1950s, the difference between the expected properties of the RNA involved in protein synthesis and those of microsomal RNA began to shake the fragile edifice of molecular biology and to cause even its most ardent supporters to question the idea of a genetic code. In a letter published in *Nature* on July 12, 1958, the Russians Andrei Belozersky and Alexander Spirin reported a chromatographic analysis of the base composition of DNA and RNA from nineteen different species of bacteria. The composition of DNA showed strong interspecific variation, but the RNA tended to be remarkably constant from species to species. According to the letter, "The greater part of the ribonucleic acid of the cell appeared to be independent of the deoxyribonucleic acid."[19]

As was the case with Avery's discovery, several major problems prevented this work from going forward. Analyzing them will highlight the important changes that had taken place during the early development of molecular biology.

A major problem was the role attributed to genes. As noted, classical geneticists had a view of the role of genes different from that of molecular biologists. The former gave the gene a distant control function whereas the latter believed that the gene determines protein structure down to the finest detail.

At the end of the 1950s, the classic conception of the role of genes was still dominant. Inherited from genetics, it agreed with what little biochemical data existed at the time. For example, Jean Brachet had shown that protein synthesis continued for hours, if not days, in anucleate cells. The fact that in higher organisms genes are physically separated from the cytoplasm also tended to reinforce this idea that genes intervened in a decisive, but parsimonious, manner. Many scientists thought that the cytoplasm also contained self-replicative particles.

This hypothesis—that some cytoplasmic components had genetic continuity and were distinct from nuclear genes although related to them—had haunted biologists (especially in Germany) from the beginning of the century, and continued to do so in the 1960s. Indeed, in some respects history has proved it right: organelles such as mitochondria and chloroplasts do contain their own DNA molecules.

But at the time, the plasmagenes were given an even more fundamental role: scientists believed that most of the events of cell differentiation—the morphological modification of cells—generally depended on paragenetic cytoplasmic components and not on nuclear genes. A whole series of experimental results was more or less strictly interpreted in the light of this plasmagene hypothesis: André Lwoff's studies of the morphogenesis of ciliates, Tracy Sonneborn's work on paramecia, Sol Spiegelman's investigation of adaptive enzymes, Jean Brachet's data on cytoplasmic RNA, and even the first experiments on microsomes by Albert Claude and George Palade.[20]

The plasmagene hypothesis filled a gap in genetic theory—its inability to *explain* morphogenesis and embryonic development. This gap would begin to be bridged only with Jacob and Monod's work and their models of regulation (see Chapter 14). Plasmagene theory was also a refuge for the opponents of genetics—those scientists who thought that the cytoplasm played an active role in morphogenesis, whereas the external medium played an important role in evolution.[21]

One of the key accomplishments of the group at the Pasteur Institute was to bring genes and proteins closer together, to show that genes intervened *permanently and directly* in protein synthesis. Genes were no longer seen as isolated particles; despite their metabolic stability, they actively participated in the life of the cell.

The "distant" conception of the gene explains why it took so long for researchers to address the question of the precise structural relation between DNA and RNA. In 1959 Crick was still speaking of the translation of DNA into RNA and seemed to be equating two processes that today are thought to be completely different.[22] As seen earlier, he also argued that RNA "controls" DNA—a vague term that merely reflected the vagueness of his idea. Even more curiously, a number of biochemical studies had begun to show that RNA was synthesized only in the presence of DNA.[23] Today it seems quite obvious that these studies showed that information is transferred from DNA to RNA. But the articles barely mention any potential biological significance of these phenomena. They do not even discuss the problem of protein synthesis: DNA seems to have merely had the function of "stimulating" the production of RNA. Indeed, these articles never dealt with the base composition or sequence of the newly synthesized RNA.

Furthermore, despite the fact that researchers had known since 1952 that RNA could coil into a helix like DNA,[24] prior to 1959 there is no mention in the literature of the idea that DNA and RNA could form a double heterohelix.[25] The fact that this hypothesis was never suggested shows that the synthesis of RNA on the basis of DNA was not considered analogous to the replication of DNA. Even in the first articles that showed the existence of short-lived messenger RNA,[26] the only data that suggested that these RNA molecules were derived from DNA were measures of base composition. There was no direct proof that RNA was an exact copy of DNA.

The first clear description of the idea that DNA was copied into RNA can be found in a 1959 article published by Mahlon Hoagland in *Scientific American.*[27] Even though Hoagland confused messenger RNA and ribosomal RNA, this article contains the idea of complementarity between the base sequence of DNA and the base sequence of RNA. The first experiment proving that RNA is complementary to DNA was carried out by Sol Spiegelman at the end of 1960. He showed that during infection of *E. coli* by bacteriophage T2, the synthesized molecules were complementary to the phage DNA.[28] This experiment closely followed Paul Doty and J. Marmur's demonstration that the two strands of DNA, after separation by heating, could re-anneal if cooled slowly enough.[29]

Why did it take so long to make this link between DNA and RNA? Why can no trace of it be found in the scientific literature prior to 1959? The simplest explanation would be that the link was known and that it was obvious to all concerned that RNA was an exact copy of DNA. But this was not so: the confusion in the articles of the time reflected the uncertainty of the scientific community. The most curious point, however, is not so much that people did not realize the exact relation between DNA and RNA, but rather that no one appeared to be interested in the question. Researchers seemed satisfied with phrases like "DNA controls the fabrication of RNA."

Before the relations among DNA, RNA, and proteins could be precisely understood, it was necessary to abandon definitively the notions of specificity and of a "template." Although it might seem that the death sentence of these ideas was signed by Watson and Crick in 1953, when they first put forward the idea of a genetic code (see Chapter 12), this was not the case. In fact, the first code—proposed by Gamow—implied a precise structural relation between DNA and the amino acids that were coded by the DNA.

Following from Gamow's model, a number of other genetic codes were proposed, none of which had anything to say about the structural relation between nucleic acids and proteins. The adaptor hypothesis, put forward by Crick in 1954, explicitly excluded all direct interaction between bases and amino acids. It would, however, be wrong to conclude that the idea of form had completely disappeared from the minds of biochemists and molecular biologists: the substantial attention devoted between 1950 and

1960 to the three-dimensional structure of RNA in the hope that this structure would reveal the mechanisms of protein synthesis (Chapter 11) shows that this was not the case. Without doubt, the final avatar was the importance given to microsomal particles, which were seen as the "workbench" on which proteins were "formed."

By giving microsomal particles a secondary role in protein synthesis, the work of the French school represented a final break between form and information and made it possible at last for molecular biology to come of age.

CHAPTER 14

The French School

In 1965 François Jacob, André Lwoff, and Jacques Monod received the Nobel Prize for Medicine or Physiology for their research on the mechanisms of gene regulation in micro-organisms. Researchers had already determined the chemical nature of the gene, the molecular structure of DNA, and the correspondence between genes and proteins—the genetic code. The work of Jacob, Lwoff, and Monod provided the final step in the circle of information exchange within organisms. They explained how regulatory proteins controlled gene expression—that is, the synthesis of the protein coded for by the gene—by binding to the gene itself.

The work of the French group was universally praised for its elegance and its association of biochemical techniques with the most advanced tools of bacterial genetics.[1] The group's results opened the way to the understanding of embryonic development in higher organisms. The study of biological deregulation (for example, cancer) also benefited from their work.

The success of the Pasteur Institute group was the result of an unexpected convergence between Monod's biochemical approach and the genetic approach of Lwoff and Jacob.[2] The origins of their research can be found in the specifically French tradition followed at the Pasteur Institute.

Jacques Monod was born into an old French Protestant family. He began his research career at the Sorbonne, where he worked on the nutritive requirements of micro-organisms.[3] In 1936, together with Boris Ephrussi, he visited Morgan's laboratory in the United States, where he learned about genetics. During that trip, he also became aware of the "backwardness" of French biology, which was partly due to France's over-rigid university system.

During his research at the Sorbonne, Monod discovered and characterized the phenomenon of "diauxy": when two food sources were added to a microbial culture, the bacteria ate first one type of food, then, after a latency period, the other. This effect produced biphasic growth curves. André Lwoff was the first to interpret the phenomenon correctly. In so doing he pushed Monod onto the road he was to follow for more than twenty years.[4] Lwoff realized that Monod's observation was an example of "enzymatic adaptation," a phenomenon discovered in 1900 and given its name in 1930 by the Finnish biologist Henning Karström. When microorganisms are put in the presence of a new food source, they are able to synthesize the enzymes they need to use the new food. The system Monod chose was that of adaptation to lactose, which led to the synthesis of the degrading enzyme for this sugar, β-galactosidase.

These adaptive enzymes were a model for studying protein synthesis.[5] The experimenter could easily induce the synthesis of an enzyme by adding a simple chemical compound to the culture medium. Such systems appeared to be ideal for determining the relative roles of genes and environment in the biosynthesis of proteins and enzymes.

As seen earlier, Beadle and Tatum's work had confirmed that enzyme synthesis was under genetic control. In the case of adaptive enzymes, the added food source—lactose in Monod's system—also played a key role. The simplest explanation of the experimental data was presented implicitly in the various articles written by Monod after his arrival at the institute (see Chapter 12). According to this view, under the action of genes, adaptive enzymes were synthesized in the form of inactive protein precursors.[6] The inducer—lactose—formed a stereospecific complex with this protein precursor and converted it into a molecule of the active enzyme, β-galactosidase.[7] Developments in the decade that followed led Monod to question the assumptions of this model and to approach the problem of adaptation from a completely new angle.[8]

Monod's initial studies appeared to confirm the model. In 1952 he succeeded in purifying β-galactosidase and obtaining antibodies against it. With these antibodies he was able to show that the precursor—the Pz protein—was present before the addition of the inducer and appeared to diminish following induction. The next year, 1953, the problems began. Experiments using a radio-labeled amino acid that was a component of β-galactosidase clearly showed that this enzyme was synthesized *ex nihilo* after addition of the inducer. The Pz protein disappeared into the ether of failed experiments,[9] and even its name was expunged from subsequent accounts.[10]

At around the same time, Monod and his colleagues separated the inducing function and the substrate function: a chemical analogue of lactose could be an excellent inducer without being metabolized by β-galactosidase. The existence of these "free" inducers raised fundamental questions about the adaptive value of the system. In a letter sent to *Nature* in 1953, Monod and the other key researchers who were studying these adaptive enzymes proposed to change their name to "inducible enzymes,"[11] a term that described the effect without ascribing any finality to it.

Such brusque changes of vocabulary are extremely rare in science, especially when they involve a very common term. Of Monod's six key articles published between 1944 and 1947, four contained the word "adaptation" in the title. Scientists generally admit that their vocabulary is not entirely precise, and that it may sometimes hide confusion. Far rarer are the cases where the dropping of a scientific term is decided *ex cathedra*.

This change of name was far from neutral: Monod's motivations were fundamentally political. By changing the name he wanted to indicate that his research was not part of a neo-Lamarckian current and even less so was it part of that contemporary version represented by the Soviet agronomist T. D. Lysenko. Monod thus spectacularly crowned his break with the French Communist Party, which had called on intellectuals and scientists to support the new Soviet theories.[12]

In the years that followed, Monod showed that other proteins besides β-galactosidase were induced by the same inducer (lactose): one of them, lactose permease—studied by Georges Cohen—enabled the sugar to penetrate the bacterium. Monod had obtained mutant bacteria, some of which showed altered functions of β-galactosidase or of lactose permease, whereas others showed changes in induction itself. Some of these mutants exhibited a constitutive expression of β-galactosidase and lactose permease in the absence of an inducer. Monod did not take the characterization and localization of these mutations very far because they required complex genetic techniques that could not be performed in his laboratory. He therefore began a joint project with Jacob. This work quickly became a close intellectual collaboration. The first experiments showed the similarity of the systems that the two groups had previously studied separately.

Jacob's research was part of an older and more "Pasteurian" project. Bacteriophages had been discovered in 1915 by Felix d'Hérelle at the Pasteur Institute and by Frederick Twort in Great Britain. Their discovery led to the hope that it would be possible to fight pathogenic bacteria by using specific bacteriophages. These initial hopes were soon dashed, however. In 1925 Jules Bordet of the Brussels Pasteur Institute, together with Oskar Bail, discovered the phenomenon of "lysogeny":[13] some bacterial strains were resistant to the destructive action of bacteriophages. But within these strains some bacteria would spontaneously lyse and release bacteriophages into the medium. These "lysogenic" bacteria thus probably contained an inactive form of the bacteriophage.

Although lysogeny had been widely studied in France during the interwar years, in particular at the Pasteur Institute by Eugène and Elisabeth Wollman, it had not attracted the attention of the other groups working on bacteriophages[14] (Delbrück thought that the observations on lysogeny were worthless). The phenomenon was uncontrollable since the induction of lysogenic bacteria took place spontaneously.

Partly in reply to Delbrück's criticisms, Lwoff began work on lysogeny after the end of the Second World War. Using the micromanipulator developed by Pierre de Fonbrune at the Pasteur Institute and a culture of lysogenic bacteria, Lwoff isolated individual bacteria that he observed undergoing successive divisions. He was thus able to demonstrate that "lysogenic power"—the ability to release bacteriophages—was transmitted, implying that the bacteriophage was present in the bacteria in a cryptic form, which Lwoff called the "prophage." Lwoff was able to induce this prophage in a reproducible way by irradiating the bacteria with ultraviolet light. This discovery was of fundamental importance because it made the study of lysogeny much easier. It also led Lwoff to recruit François Jacob, a young doctor who had recently come out of the army knowing nothing of biological research, but who had been inspired by reading Schrödinger's *What Is Life?*[15] In the space of a few years, using all the resources of bacterial genetics, François Jacob and Elie Wollmann gradually revealed the complex mechanism of lysogeny.[16]

They showed that, at the prophage stage, the *E. coli* bacteriophage called λ (lambda) was closely associated with the bacterial chromosome. Using strains of bacteria that showed high rates of conjugation (Hfr strains) and a Waring blender (the same kind of mixer used in the Hershey-Chase experiments) to interrupt conjugation at different times, they were able to localize precisely the prophage on the chromosome (see Chapter 5). Strangely enough, the bacteriophage was induced during conjugation, as soon as it entered into the bacterium—Jacob and Wollman called this phenomenon "erotic" induction. But lysogenic bacteria that carried a prophage were resistant, "immune" to infection by other bacteriophages. This immunity was the result of the action of a single dominant bacteriophage gene—the C locus. These data remained unexplained until one of the first results of the collaboration between Jacob and Monod put them in a new light.

Jacob and Monod's work, carried out in a loft in an old building of the Pasteur Institute (hence the nickname of the lab—the *grenier*), soon showed that the same principles of regulation were at work in the two systems. The generality of the phenomenon made Jacob and Monod's discovery particularly important. Remarkably, both groups used the term "induction" without considering this coincidence in any way significant.[17] In fact, Lwoff formed the living link between the two research projects, both of which dealt with phenomena that appeared inexplicable from the point of view of strict genetic determinism. As will be seen, this was a unique characteristic of French biology in the 1930s to 1950s.

The first experiments were simply intended to apply the technique of conjugation to the study of β-galactosidase induction. These "PaJaMo" (or more often "pyjama") experiments produced some surprising results.[18] When a male bacterium that could produce β-galactosidase only in the presence of an inducer (lactose) was crossed to a mutated female bacterium that synthesized an abnormal form of the enzyme, normal β-galactosidase was rapidly produced in the absence of the inducer. This synthesis was transitory and would cease after a few hours.

These experiments showed that the action of the regulator gene required the formation of a cytoplasmic product that must inhibit β-galactosidase synthesis. The product of this regulator gene was called the "repressor." The induction of β-galactosidase during mating was analogous to the "erotic" induction of the bacteriophage λ described above: the same mechanisms should explain lysogeny and the control of inducible enzymes. Monod and Jacob later showed that the products of regulator genes—repressors—were proteins. Naturally active, these repressors were inhibited by lactose (or, more precisely, by a substance derived from lactose) and, in the case of the bacteriophage, by UV light. For a while, the mechanism of action remained unclear: did repressors act at the level of the gene, by controlling its transformation into RNA, or at the level of protein synthesis? Jacob and Monod opted for the first hypothesis: repressors blocked the copy or transcription of genes into RNA[19] by binding to the DNA sequences, called operators, situated upstream from the regulated genes.

Jacob and Monod's experiments also showed that as soon as the repression of a gene stopped, protein synthesis was quickly induced. Similarly, once a gene was inactivated, synthesis suddenly came to an end. This implied that there must be a short-lived intermediary between genes and proteins that Jacob and Monod later called messenger RNA (see Chapter 13).

Few experiments have been so rich and have led to so many discoveries as the PaJaMo experiment: the discovery of messenger RNA, the proof of the existence of repressors, and the elaboration of a general schema of negative regulation of gene expression. But who should be credited with these discoveries, Jacob or Monod? Following his observation of "erotic" induction, Jacob was certainly the better placed to interpret the results of the PaJaMo experiment. Furthermore, Monod had not completely abandoned the idea that the inducer played a role in protein folding and in the conformation of inducible enzymes.[20] He was thus not prepared to accept that the inducer acted on a protein—the repressor—that was distinct from the inducible proteins β-galactosidase and lactose permease.

The analogy of the erotic induction of the prophage and the transitory activation of β-galactosidase in the PaJaMo experiment was clearly stated in Jacob's Harvey Lecture, given in the summer of 1958.[21] In order to explain how the phage repressor could simultaneously inactivate *all* viral functions, Jacob put forward the hypothesis that the repressors had to act directly on DNA. This hypothesis was particularly problematic for those such as Monod who thought that genes were distant, isolated, and untouchable structures.

These different views reflected the different systems being studied. Bacterial geneticists like Jacob had come close to the gene and had even begun to "fragment" it. The "classical" geneticists and other biologists, such as Monod, had retained a much less "realistic" view of the nature and role of genes. If Jacob played an essential role in the development of the regulation model, Monod was key to the elaboration of the model, to the definition of its different components, and to the design of experiments that would prove its validity in an extremely elegant manner.

Jacob and Monod confirmed that several structural genes could be controlled by a single regulator gene. These genes were generally grouped together in a structure called an "operon" and were transcribed into a single messenger RNA from a special DNA sequence called a promoter. Jacob and Monod's model of regulation was thus known as the "operon model."

One characteristic of this model, the idea of a repressor, requires closer historical analysis. It was several months before Monod accepted that induction could be caused by the inhibition of repression, that is, that a *positive* phenomenon could be caused by a *double negative.* For a long time he had considered induction—adaptation—a positive phenomenon, the result of a complex interaction between the cell and the perturbation that was imposed upon it. This kept him from accepting the inducer's minimal role in the new model. It was only after a seminar given by Leo Szilard that Monod was convinced that regulator genes coded for repressors.[22]

After the war, Szilard had abandoned his research in physics and begun to study bacterial metabolism, helping to discover the inhibition of metabolic pathways by a feedback effect of the end-product. The characterization of the mechanisms involved was one of the most effective applications of cybernetics to biology[23] and led to the regulatory vision of molecular biology and to the study of "allosteric" enzymes. This work put Szilard into contact with Monod and led him to play an active role in developing the conception of induction as a de-repression. Szilard had an unusual career, abandoning a subject as soon as it had been even partly resolved. Like Delbrück, he became interested toward the end of his life in what seemed to be the final frontier—the study of the nervous system—and proposed a number of models to explain memory. An ideas man rather than an experimentalist, Szilard, like many of his physicist colleagues, played an essential role in the clarification of biological problems.[24]

In addition to genes that code for enzymes or structural proteins, there are regulator genes, whose sole function is to control the activity of other genes. The impact of this fundamental idea on the development of molecular biology has not been sufficiently emphasized.[25] It is particularly interesting to note that the rare experimental data available to Jacob and Monod did not justify either this clear distinction between structural genes and regulator genes, or the generalization of the concept of a regulator gene.

According to Jacob and Monod, this distinction, which was clearly set out in an article published in the *Comptes Rendus de l'Académie des Sciences* in October 1959,[26] rapidly became the "first postulate" of their models of gene regulation.[27] It appears that it was Jacob who, inspired by studies of the bacteriophage λ that showed the importance of a single gene in the control of lysogeny, first conceived of the distinction between regulator genes and structural genes.

This distinction introduced a hierarchy of genes. It also changed the aim of research in molecular biology. For the molecular biologists of the 1960s, one of the major objectives was to understand the functioning and development of higher organisms. How, on the basis of a single cell, could the two hundred different types of cells required to form a higher organism (such as a human) possibly develop? What mechanisms were involved in making skin or muscle cells, which possess the same genes but express these genes differentially and thus synthesize different proteins? The discovery of regulator genes implied that the work of molecular biologists should center on the study of these genes, and that this would be both necessary and sufficient to understand the development of higher organisms.

The distinction made by Jacob and Monod thus defined a precise, but difficult, research program: characterize regulator genes in higher organisms and understand how these genes interact and are organized in a network. The early years of this work were particularly unrewarding: the products of these "master genes"—regulatory proteins—were so rare that they could not be detected. The first regulator genes were isolated only by the use of genetic engineering. This was facilitated by the fact that these genes have been conserved in the course of evolution.

The idea of the regulator gene also played an important role in cancer research—a cancerous cell is a cell in which gene expression is deregulated (Chapter 19)[28]—and, to a lesser extent, in studies of molecular evolution (see Chapter 21).

The concept of the regulator gene is important not only because it oriented molecular biological research programs for more than thirty years, but also because it closed—virtually definitively—another line of research that had been partly opened by the German-born geneticist Richard Goldschmidt and had been developed in particular by the American geneticist Barbara McClintock's work on maize.[29] According to McClintock, the differential expression of genes resulted from their physical movement in the genome. As they moved, genes were placed under the control of different regulator elements that modulated their expression. This genetic "transposition" was under the control of another gene. McClintock had in fact anticipated the discovery of regulator genes, but the complexity of the system she studied, difficulties in explaining her results to other scientists, and above all the key importance she gave to the physical movement of genes in regulation all limited the impact of her findings.

The work of François Jacob and Elie Wollman on the bacteriophage and conjugation showed that the movement of genes in bacteria could also alter gene function. Two models of genetic regulation were thus possible: one saw this regulation in terms of a stable genome, under the control of networks of regulator genes; the other linked the regulation of gene expression to a structural modification of the genome that took place during the lifetime of the organism.

In June 1960 Jacob published an article in *Cancer Research* in which he explained to cancer specialists the mechanisms of regulation that existed in micro-organisms, placing the two possible forms of regulation on an equal footing.[30] In the book he had published with Elie Wollman a year earlier *(La sexualité des bactéries),* Jacob argued that mobile genetic elements, which he called episomes, could intervene during cell differentiation to modify nuclear potential.[31] In the conclusion to the 1961 Cold Spring Harbor Symposium,[32] then more clearly in a 1963 article (written with Monod), Jacob definitively opted for a model of stable regulation with no physical alteration of the genome: "most of the known facts support the view that the genetic potentialities of differentiated cells have not been fundamentally altered, lost, or distributed" (p. 31).[33] (Jacob's choice might have reflected the influence of Monod, who, considering nature straightforward and Cartesian, was always inclined to chose between two competing hypotheses.)

The notion of regulator genes and the distinction between them and structural genes had major scientific consequences. This functional separation also had a historical and philosophical importance. It brought genetics closer to embryology and raised the possibility that it would be merely a matter of time and hard work before they fused definitively. Philosophically speaking, it represented the final step in the development of a new vision of biology that had begun in the 1940s and that made the question of the information contained in genes an ordering principle of all life. Jacob and Monod went further, distinguishing two kinds of information in organisms: structural information that was necessary for the formation of the components of the organism, and regulatory information that was responsible for the gradual spatial organization of these structural components during development. This distinction had been outlined by Schrödinger in *What Is Life?* and found its clearest expression in the work of Jacob and Monod. Edward Yoxen has shown that Schrödinger's book, which had been very influential in the early days of molecular biology but had subsequently been forgotten, was rediscovered in the 1960s.[34] Jacob and Monod's articles contain formulations similar to those used by Schrödinger;[35] the question remains whether this was the result of a direct influence[36] or an indication of the continuing relevance of Schrödinger's ideas and of a strange resonance between his speculations and the models developed by the molecular biologists.

After more than five years of joint work, Jacob and Monod went their separate ways. Monod returned to the study of a class of proteins—regulatory proteins—for which the repressor was the model. The repressor protein interacts with DNA, but that interaction may be inhibited by a breakdown product of lactose binding at another site on the repressor molecule. Many other proteins or enzymes, and in particular those subject to negative feedback control, have similar properties: their activity can be modulated by activators or inhibitors binding to sites other than the catalytic site.

Monod, together with Jean-Pierre Changeux and Jeffries Wyman, developed the "allosteric model," which explained the properties of these proteins and enzymes.[37] This model had largely been inspired by work on hemoglobin, from Linus Pauling's pioneer studies in the 1930s to the most recent results from crystallography.[38] Other models of regulation were proposed by competing groups, and for several years there was a lively debate over allosteric theory.[39] Time—and the structural data obtained by X-ray diffraction—generally supported Monod's allosteric theory, although it did not turn out to have the universality that Monod had hoped for.[40]

In retrospect, two characteristics of the allosteric model are particularly striking. The first is that its dogmatic presentation, in the form of a series of postulates, was over-rigid and in some cases unnecessary.[41] This kind of dogmatism was probably reinforced by the absence of a strong tradition of structural chemistry in France. These postulates, however, were extremely rich in terms of the information they provided about the enzyme and its structure. The other models, which were much less restrictive, described the kinetic behavior of enzymes without seeking to explain their structural bases.

The second characteristic of the model that distinguished it from its competitors was linked to the first: whereas the other models were almost always able to describe the behavior of enzymes, the allosteric model made a number of easily falsifiable predictions.

These two characteristics of the allosteric model were the result of Monod's persistence in trying to impose some kind of order on the anarchy of life. They were also the product of his adoption of Karl Popper's philosophy of science,[42] according to which science advances through easily falsifiable models and not through theories that explain everything. In the context of contemporary molecular biology, allosteric theory appeared as an island of rigid Cartesianism in a sea of Anglo-Saxon pragmatism.

François Jacob remained faithful to the research program outlined in the 1963 article; he studied increasingly complex regulatory circuits before turning to the regulation of gene expression in higher organisms. Continuing his work on the bacteriophage λ, Jacob and his coworkers described complex regulatory circuits that controlled the transition between infection and lysogeny. They then went on to study the regulation of cell division, an essential step in the bacterial life cycle. Division takes place only when the bacteria have become large enough and when there is sufficient food available. Jacob, Brenner, and Cuzin developed a regulator model—the "replicon"[43]—which more than thirty years later still serves as a guide for researchers. Finally, Jacob bridged the gap that separates bacteria from higher organisms, and in 1970 began to study the embryonic development of the mouse.[44]

Jacob and Monod's research put the finishing touches on molecular biology and laid the foundations of future molecular biological research programs. But French researchers had been virtually absent from the early stages of the development of this new science. Furthermore, the two roots of molecular biology—biochemistry and genetics—developed much later in France than in other countries.

There is a striking contrast between the brilliant interwar developments in biochemistry that took place in Germany, then in Britain and America, and the virtual nonexistence of French biochemistry in the same period. This delay can mainly be explained by the actions of one man, Gabriel Bertrand,[45] the chair of biological chemistry at the University of Paris and the director of one of the main biochemistry laboratories at the Pasteur Institute. After a series of striking studies on enzymes, Bertrand decided that proteins played no real part in them and that the key catalytic role was played by the metal ions that are often found associated with enzymes. The orientation of Bertrand's research and its decisive influence meant that French biochemistry played no part in the subsequent characterization of the various major metabolic cycles or in the study of proteins, despite the medical and agronomic importance of Bertrand's research.

The reasons for France's backwardness in genetics are more complicated. Some historians have argued that the slow rate at which chairs in genetics were created in French universities was a result of the top-heavy and over-rigid French university system, a sign of its "fossilization." Other reasons can equally well explain the delay: at the end of the nineteenth century France was particularly slow in accepting—or even discussing—Darwin's theory,[46] and in the first half of the twentieth century neo-Lamarckism was very strong among French biologists.[47] They did not directly reject the results of Mendelian genetics, but they nevertheless tended to minimize the role of genes in the functioning and development of organisms while emphasizing the role of the cytoplasm.[48]

To understand how, despite these apparent handicaps, France nevertheless played a major role in the development of molecular biology, one must see this weakness of biochemistry and classical genetics as an advantage for the French researchers.[49] As shown earlier, molecular biology often developed *against* biochemistry, by showing that proteins and enzymes were not at the heart of self-replicative phenomena, by opposing the idea that biological facts could be explained by thermodynamic rules linked to the totality of chemical reactions that take place in the organism. The

French molecular biologists were freer than their Anglo-Saxon colleagues because they spent less of their time fighting the biochemists.

Relations with French geneticists were both more complex and more productive than those with the biochemists. The genetics laboratories that were set up in France after World War Two, such as Boris Ephrussi's laboratory at Gif-sur-Yvette, had an unusual approach to the subject.[50] Turning away from the characterization of chromosomal genes, which was dominant in the U.S. school of genetics, the French tended to concentrate on the study of hereditary phenomena that were apparently independent of the nucleus. This approach was the continuation of a tradition begun with André Lwoff's work on the genetic continuity of infracellular organelles in ciliates—unicellular organisms[51]—and with the work of Elisabeth and Eugène Wollman on the fate of phage in lysogenic bacteria. To this concentration on paragenetic phenomena one must add the influence of Claude Bernard. As Jean Gayon and Richard Burian have pointed out,[52] his approach emphasized physiology rather than morphology. For Bernard, organisms were characterized by their ability to adapt continually to physiological changes rather than by their structure or form. Whereas the geneticists of Morgan's group were interested in the genes controlling the form of organisms (the wing-structure of flies, for example), the French groups were interested in the genetic control of organic adaptability.[53] They were able to show that an adaptive process, conditioned by the external medium, is nevertheless subject to strict genetic control.

But more than any of these French particularities, it was the very special framework of the Pasteur Institute that offered Lwoff, Jacob, and Monod a favorable environment for their research. The institute allowed the various groups that worked there complete scientific autonomy and total independence from the university system. Despite the lethargy that had overtaken the institute in the first half of the century, during the long directorship of Emile Roux, its prestige remained intact. The institute continued to attract large numbers of visiting foreign scientists from both Europe and America. This independence and prestige enabled several groups, including Lwoff's, to be fully integrated into the burgeoning international network of molecular biology.

PART III

The Expansion of Molecular Biology

CHAPTER 15

Normal Science

Once researchers had deciphered the genetic code and described regulatory mechanisms in micro-organisms, molecular biology entered what the historian of science Thomas Kuhn has called a period of "normal science."[1] Research no longer involved testing global models but "puzzle-solving" within the framework of existing theories. Molecular biologists did not think they had solved all the mysteries of biology, but their understanding of fundamental molecular mechanisms appeared sufficient to imagine how the unresolved problems (morphogenesis, the origin of life) might be approached. This impression of completeness was shared by most molecular biologists (as it had been by physicists in the 1930s),[2] and it led some of them to turn to what appeared to be the final frontier of human knowledge—the nervous system.

In the 1950s Max Delbrück, the founder of the phage group, had been disappointed that the study of genes had not revealed new physical principles, and had turned instead to the study of a fungus that is sensitive to light and other stimuli as a model for studying sensory physiology at the molecular level.[3] Seymour Benzer, a specialist in bacteriophage genetics, turned to *Drosophila* behavior genetics.[4]

Other molecular biologists, such as Gunther Stent and François

Jacob, wrote the history of the subject, reinforcing the impression that a key phase in the discipline's history had come to an end.[5] Finally, in *Chance and Necessity,* Monod set out the principles of the new ethics that molecular biology seemed to elicit. The book was the consequence of the new view of the biological world and an eyewitness account of its development:

> The theory of the genetic code constitutes the fundamental basis of biology. This does not mean, of course, that the complex structures and functions of organisms can be deduced from it, nor even that they are always directly analyzable on the molecular level. (Nor can everything in chemistry be predicted or resolved by means of the quantum theory which, without question, underlies all chemistry.) But although the molecular theory of the genetic code cannot now—and will doubtless never be able to—predict and resolve the whole of the biosphere, it does today constitute a general theory of living systems.[6]

Monod's book brought the principles of the new science to the attention of the general public in France and elsewhere and led to one of the last great French intellectual debates of the second half of the century. Dozens of books were written, all replying in slightly different ways to Monod's views.[7] The heart of the debate was the role of chance in evolution, and thus the place left for a religious conception of the universe by the new scientific view.[8]

This conviction that the essentials of the new science had been discovered inevitably led to dogmatism and a refusal to accept any results that tended to question—no matter how slightly—what was considered already proven. Two examples will suffice to illustrate how molecular biology was affected by such narrow-minded reactions. The first is taken from the work of Jacques Monod and his colleagues on regulation in microorganisms. Chapter 14 showed that Monod found it difficult to accept the existence of negative feedback control. Other systems of bacterial regulation, such as that studied in Monod's laboratory by Maxime Schwartz and Maurice Hofnung, did not follow the model of negative regulation. Other groups had already proposed models of positive regulation, which Monod rejected. It was only after a great deal of effort that Schwartz and Hofnung were able to convince Monod that positive regulation did indeed exist.[9] In fact, this concept was to be of great importance in bacteria and especially in higher organisms (see Chapter 18).

The other example of the resistance of molecular biologists to even a partial questioning of their results was the very important discovery of an enzyme that could copy RNA into DNA. The existence of such an enzyme had been proposed by Howard Temin in 1962, but was only accepted eight years later.[10]

To understand the importance of this enzyme and the questions raised by its discovery, it is necessary to return to Francis Crick's declaration of the "central dogma" in 1957, in his famous lecture to the Society of Experimental Biology (see Chapter 13).[11] This lecture was important because, for the first time, Crick gave a synthetic presentation of the principles of the new science. He discussed the nature of the genetic code and proposed the adaptor hypothesis. Most important, he set out the "philosophy" of the new science: all the information required to create organisms can be found in molecules of DNA. This information goes into proteins, where it determines the order of amino acids. No transfer of information in the other direction—from proteins to DNA—is possible. On the one hand, no part of the cellular machinery appeared able to convert the three-dimensional information contained in proteins into the one-dimensional information contained in a DNA molecule. On the other hand, the existence of such a transfer would call into question the fundamental principles of genetics. Allowing the possibility of a transfer from proteins to DNA would open the door to neo-Lamarckism; indeed, the study of regulation in micro-organisms had shown that the medium could have an effect on protein synthesis. If proteins could change the message contained in DNA, then the environment could change DNA, and the inheritance of acquired characters would be possible. The rejection of any transfer of information from proteins to DNA was thus written in stone—more because of the traditions inherited from genetics and neo-Darwinism than because of the biochemical data that were available to Crick and his contemporaries.

As noted, Crick was particularly vague about the role of RNA. The hypothesis of messenger RNA and the experimental proof of its existence were three years in the future. Many experiments had suggested that RNA might be the precursor of proteins and be derived from genes (DNA), but the possibility of a transfer from RNA to DNA was by no means excluded; some models even proposed that RNA might be a common precursor of both DNA and proteins.

The totality of relations between biological macromolecules formed what Crick somewhat unfortunately called the "central dogma." Defending himself later, Crick explained that he was unaware of the theological meaning of the word "dogma" and that he had confused it with "axiom."[12] After the discovery of messenger RNA, the central dogma took on its definitive form: DNA was copied into RNA, which itself was translated into proteins. The passage RNA → DNA was removed from the models, without any justification beyond the fact that this made things simpler.

In science in general, and in molecular biology in particular, models have their own weight. So it was that the passage from RNA to DNA was considered to be quite simply impossible. But Howard Temin's experiments on retroviruses soon necessitated a reassessment of the dogma. Retroviruses like SV40, the adenovirus, or the polyomavirus are tumor viruses (oncogenic viruses). They are different from the other tumor viruses, however, in that their genetic material is RNA. By the end of the 1960s, the study of oncogenic viruses had become a major research topic in molecular biology. Many scientists thought that oncogenic viruses would provide a royal road to the understanding of higher organisms, much as the study of bacteriophages had helped in the understanding of bacteria.

Renato Dulbecco[13] played a very important role in developing quantitative methods for studying animal viruses. He had worked with Luria, then with Delbrück, on the rather strange phenomenon of the reactivation by visible light of phages that had been inactivated by ultra-violet light. In 1950 Max Delbrück asked Dulbecco to interrupt this work and visit a series of U.S. laboratories that were working on cell culture and the multiplication of viruses in these cultures. Thanks to the work of John

Enders, Thomas Weller, Frederick Robins, and Wilton Earle, cell culture technique had recently been improved by a series of empirical alterations (use of antibiotics, changes in medium composition) to the extent that it had become a powerful tool for the study and production of viruses. On his return to Caltech, it took Delbucco only a few months' work with the horse encephalitis virus to develop a "lysis plaque technique"—similar to that used to detect bacteriophages—which made possible a quantitative measurement of viruses.

Delbucco first applied this new approach to the poliomyelitis virus; then in 1960 he turned to the study of a small oncogenic DNA retrovirus, polyomavirus. Instead of the plaques of lysed cells, plaques of transformed cells appeared on the surface of the dish.

Dulbecco's work shows that Delbrück's influence extended far beyond the world of the phage. It also shows how molecular biologists were able to profit from the new technological tools being developed in other areas of biology. But above all it provides a very concrete example of the influence of research funding on scientific developments. Delbrück's approach can be explained by two very large grants he received, one from the National Foundation for Infantile Paralysis (set up by President Roosevelt to support research into polio, it provided substantial grant aid to virology laboratories), the other a sizable donation to Caltech by the American millionaire James G. Boswell, who suffered from shingles.

Studies of small oncogenic viruses had shown that their DNA was integrated into the genome of infected cells and that this process was linked to the "transformation" of the cells—their passage from a normal to a cancerous state. Temin had previously proposed the same hypothesis to explain transformation by retroviruses. In his model, Temin was obliged to imagine the existence of a conversion step from RNA to DNA, preceding the integration of the retroviral genetic information into the genome, which he felt had been proved by the effect of different inhibitors on transformation.

Nevertheless, it was more than eight years before Temin convinced the rest of the scientific world and the central dogma was "reversed," as an editorial in *Nature* put it in 1970.[14] The scientific community was finally convinced when Temin and another group, led by David Baltimore, isolated an enzyme associated with the viral particle able to convert RNA into DNA and thus called "reverse transcriptase." There can be no doubt that this discovery was made all the more difficult because of the prevailing dogmatism of molecular biology, and by the certainty that existing results were established truths.

In hindsight, the whole controversy over the existence of reverse transcriptase seems ridiculous. After all, its discovery did not shake molecular biology to its foundations. As will be seen, it in fact provided one of the most important tools of genetic engineering, which in turn was to reinforce the models of molecular biology.

But if Temin did not destabilize molecular biology, it was not for want of trying. No sooner had reverse transcriptase been discovered in retroviruses than Temin proposed a "protovirus" model, according to which retroviruses were merely the pathological products of the normal cell machinery, which was able to copy RNA into DNA. Temin argued that the amount of DNA in different cells in the organism varied, depending on the quantity of RNA present in the cells and their functioning. He put forward the hypothesis that this reverse transcription machinery played an essential role in development,[15] in complete opposition to the models of Jacob and Monod (see Chapter 14). A reverse transcriptase–type activity in healthy, non-infected cells has still not been found, however, and eventually Temin was obliged to abandon his protovirus theory.

The period from 1965 to 1972 was characterized by an increase in the number of groups and laboratories working in molecular biology, the creation of institutes devoted to the subject, and above all an expansion of the molecular vision beyond its original field, leading to a take-over of other biological disciplines by molecular biologists.

Although this "molecularization" was generally gradual, it sometimes took place swiftly: a single article could push a whole discipline into the "molecular field." One of the best examples is that of biological membranes. Both physico-chemical and morphological (electron microscopy) studies had suggested that membranes were formed of a lipid bi-layer, one side of which was covered with proteins. In 1972, S. J. Singer and Garth L. Nicholson proposed a very different model, according to which proteins were inserted fully into the membrane.[16] This model implied that proteins played an essential role in the structural and functional properties of membranes. The study of membrane properties thus escaped the orbit of lipid biochemists and became the domain of protein specialists, and in particular of molecular biologists.

These "take-overs" took place in a number of ways. Molecular biologists received important funding, often at the expense of other researchers. They took control of scientific journals, either by creating new publications devoted to the new "molecularized" sciences,[17] or by changing the editorial line of existing journals.

Molecular biologists also took over in the universities, either by introducing molecular biology into the curricula or, more often, by "updating" biochemistry or genetics courses. Particularly important in this struggle were job descriptions in both teaching and research, and the choice of candidates. In both respects, the molecular biologists clearly came out on top. A later section outlines the results of the introduction of the new science into the main biological disciplines. Here, however, it is necessary to understand how the molecular biologists were able to take control and to impose their new vision of biology onto the whole of the biological community. This study will be limited to France,[18] where the belated introduction of molecular biology and the isolation of the pioneering groups of Lwoff, Jacob, and Monod make this seizure of power and "molecularization" of biology all the more spectacular.

For the "Whig" historian of science, none of this would make much sense. He or she would argue that molecular biology succeeded simply because of its worth, because of the beauty of its results, and because of the new areas of research that it opened up. For sociologists of science (or at least, for the most dogmatic of them), by contrast, a new science can have no "objective" superiority. Superiority is merely the product of confrontation, not the reason for victory. It is thus necessary to think in terms of the strategies of disciplines, and not of the "value" of scientific theories. For

Max Planck, a new theory is accepted only when the partisans of the old theory have died. The triumph of molecular biology, however, was far too rapid for Planck's explanation to be true!

No one can deny that molecular biology is a compelling science. The primary reason for this lies in its simple and pedagogic models. These models are not a way of popularizing the science but form one of the basic tools used by molecular biologists in their daily work. They make molecular biology attractive. Take Jacob and Monod's models of regulation: how many young biologists have been impressed for life by these models, which seem to clarify the complexity of biological phenomena? As early as 1944, the clarity and simplicity of Schrödinger's book had attracted many young physicists to biology. Twenty years later, the pedagogic virtues of Jacob and Monod's models would lure young biologists away from the traditional disciplines.

These models are persuasive because the logic that inspired them and the image of the biological world that they reveal are in harmony with the picture of the world that is presented both by the media and by other scientific disciplines. Explaining the functioning of organisms in terms of information, memory, code, message, or negative feedback involves using a language and a set of images that everyone knows.

It is clearly not the results produced by molecular biology that have convinced people of its worth. As we will see, the study of higher organisms was a lot more difficult than expected. Few new results were obtained and the onward march of knowledge proved to be difficult. The sociologists of science are at least partly right: molecular biology did not win simply on the battlefield of facts.

In France the rise to power of the molecular biologists was assisted by Jacob, Monod, and Lwoff's Nobel Prize (1965). The Nobel Prize confers enormous prestige and gives the prize-winner influence at all levels of the decision-making process in his or her field—recruitment, financing, and the construction of new laboratories. Monod, who was appointed professor at the prestigious Collège de France, then became director of the Pasteur Institute in 1971, was able to intervene directly in the development of molecular biology in France. But this political power was a consequence of a scientific judgment by the Nobel committee: in the case of Jacob and Monod, the attribution of the Nobel Prize sanctioned a discovery that the scientific community already considered to be fundamental. This is shown by the fact that Jacob and Monod were given pride of place at the 1961 Cold Spring Harbor conference, long before they received the Nobel Prize.

Jean-Paul Gaudillière has shown, however, that the intervention of Monod and the Pasteur group in the development of molecular biology pre-dated the awarding of the Nobel Prize and the national and international recognition of their work. The main force behind the development of molecular biology in France was not the CNRS or the universities but the Délégation Générale à la Recherche Scientifique et Technique (DGRST—General Commission for Scientific and Technical Research). The DGRST, under the direct control of the prime minister, was given the task of making up for France's scientific backwardness. A creation of the Gaullist regime, it was intended to contribute to the renewal of French science by cutting out the cumbersome administrative machinery of the research organizations and the university system.

From 1960 on, one of the main tasks of the DGRST was to promote molecular biology. The priority given to this discipline was the result of the personal links that existed between the molecular biologists and those in General de Gaulle's entourage who were responsible for developing the new science policy. These links had been forged in the Resistance and had been reinforced between 1955 and 1960 at a series of meetings and conferences, such as the one that took place in Caen (Normandy) and tried to reorganize French research and universities. The political contacts that made the action of the DGRST possible had thus been established long before the work of the French school of molecular biology had reached its peak.

Yet despite the money that was poured into it, both in France and elsewhere, and despite its enormous prestige, molecular biology made little progress between 1965 and 1972. Nevertheless, some of the "dogmas" were confirmed: Crick's 1957 "sequence" hypothesis, according to which the structure of a protein depends only on the chaining of its amino acids (Chapter 13, p. 139), was demonstrated by the *in vitro* total synthesis of an active enzyme—ribonuclease A—solely from amino acids.[19] The most striking development of this period was the discovery of the three-dimensional structures of a series of proteins through the use of X-ray analysis of crystals.[20] Each structure was different: there was no simple rule for the folding of polypeptide chains. The secondary structures described by Linus Pauling were indeed present, but they appeared to play only a limited role in the creation of the three-dimensional structure. Furthermore, not everybody accepted these results: many biochemists thought protein structure might have been altered by the crystallization process that was required to make the observations.

Nevertheless, the first crystallographic analyses of enzymes led to the development of models of catalysis that showed that proteins used the principles of organic chemistry "intelligently."[21] But these developments were not necessarily the result of molecular biology: they were the final step in a long road begun many years earlier with the attempts of W. M. Stanley, Linus Pauling, and J. D. Bernal to decipher the structure of biological molecules. These data did not relate to the heart of the new discipline—the mechanisms of information exchange within cells. Little progress was made in this respect. The analysis of the mechanisms of the regulation of gene expression only complicated Jacob and Monod's seductively simple model: in addition to negative control, other mechanisms, such as positive regulation, turned out to play a role.

The molecular mechanisms of the major biological processes—replication, transcription, and translation—were all clarified: after studies on the first enzyme that could replicate DNA *in vitro*, DNA polymerase I, another enzyme, DNA polymerase III, was found to be responsible *in vivo* for the replication of DNA in bacterial cells (see Chapter 20).[22] Similarly, RNA polymerase—the enzyme that copies DNA into RNA—was purified and its action described: a small protein sub-unit, the σ factor, was found to be responsible for beginning transcription by precisely positioning the RNA polymerase upstream of the gene.[23]

Researchers characterized many of the factors necessary for protein synthesis and began to understand ribosomal structure more fully. They crystallized transfer RNA molecules and began determining their three-dimensional structure.[24] The structures of messenger RNA and of viral RNA were studied using various techniques, but without any decisive breakthroughs.

Molecular biologists also determined the first nucleic acid sequences,[25] isolated and purified the first genes from bacteria and from higher organisms,[26] and chemically synthesized a gene corresponding to a transfer RNA.[27] Studies involving gene manipulation were extremely complex—as shown by the number of authors involved—and difficult because of the inappropriate technology employed.

Many molecular biologists were aware that the new frontier of their discipline was the study of higher organisms: their functioning, their development, and their pathologies. The cells of higher organisms have a nucleus that isolates DNA from the cytoplasm, where protein synthesis takes place. The existence of this barrier introduced an additional level of complexity into the transfer of information from DNA to proteins. Furthermore, the cells of higher organisms can modify their morphology and their biochemical make-up and can differentiate during development. Molecular biologists had to try to understand how the activity of genes could be stably modified. This was also an important question in pathology: many diseases are the result of an error in embryonic development, and cancer itself appears to correspond to a de-differentiation of the cell and its abnormal "reprogramming."

Many of the founders of molecular biology turned to the study of the development of complex organisms. Sydney Brenner began to study the nematode (this microscopic worm, with only a thousand cells, nevertheless possesses an organized nervous system and can even learn). After hesitating over the choice of the same animal, Jacob finally turned to the mouse. He argued that to take the next step in molecular biology, a "mouse institute"[28] had to be built. The genetics of the mouse is among the best known of all the higher organisms, and the principal steps in its development are analogous to those of humans.

In order to study complex organisms molecular biologists had to retrain: they were obliged to plunge into the complexity of development, to learn the techniques of experimental embryology and cell biology. They could begin the molecular study of these organisms only after much preparation.

Molecular studies of higher organisms proved extremely difficult. The amount of DNA present in the nucleus of a mouse cell is a thousand times more than that in a bacterium. Furthermore, because of the large number of different cells present in a higher organism, it is much more difficult to purify a mouse protein than a bacterial protein. The isolation of the proteins that regulate transcription is difficult enough in *E. coli* (the isolation of the repressors of the lactose operon and of the phage λ took more than five years);[29] it was virtually impossible in higher organisms.

By studying isolated cells rather than the many different cells in higher organisms, molecular biologists could avoid one level of complexity. This explains why Jacob, at the same time as he carried out his work on mice, also began to study cells from mouse embryonal carcinoma, which had properties similar to those of the primitive cells of the embryo. François Gros turned to the study of cell lines that could differentiate *in vitro* into muscle cells and thus reproduce one stage of embryonic development.[30]

These molecular studies of higher organisms showed that differentiated cells have an RNA content different from that of nondifferentiated cells. They thus confirmed that cellular differentiation corresponds to the modification of gene expression, and that it is controlled mostly at transcription, the point at which DNA is copied into RNA. These studies nevertheless revealed a series of new phenomena that were difficult to interpret. The genome of higher organisms turned out to contain repeated sequences of DNA, which were present in large amounts (up to 50 percent). Gene transcription produced unstable heavy molecules of RNA called HnRNA (heterogeneous nuclear RNA), only a fraction of which (around 10 percent) entered the cytoplasm to play the role of messenger RNA (see Chapter 17).[31]

Although there were no important experimental data on the molecular mechanisms of differentiation, all these new, uninterpretable results were the object of speculation and were incorporated into new models of regulation, such as that of Roy Britten and Eric Davidson.[32] These models lacked the simplicity of their predecessors.

This overinterpretation of data in order to explain the extraordinary complexity of the functioning, organization, and development of higher organisms was almost certainly a reflection of the frustration felt by the molecular biologists. The studies turned out to be long and difficult, and the road from the bacterium to the elephant was extremely hard.[33] Molecular biologists' understanding of higher organisms was so limited that the importance of minor developments was often inflated. Although understandable, this attitude was aggravated by the researchers' ignorance of the extraordinary richness and diversity of life. The naively Darwinian ("Panglossian") concepts of some molecular biologists led them to try to justify the existence of the new molecular mechanisms they discovered.

All was not bleak during the years 1965–1972, however. Of course, it took some time for researchers who had never studied anything other than *E. coli* and the bacteriophage to get used to the complexity of higher organisms. But this period was useful for developing methods for extracting RNA and DNA for the molecular study of higher organisms. During this time links were forged between specialists from different disciplines—embryologists, cell biologists, molecular biologists—informal networks were established, and different laboratories became focused on similar objects. These years were clearly necessary for the "molecularization" of biology.

During the 1960s, the use of the word "molecular" in the titles of laboratories, disciplines, and academic courses spread like wildfire. Everything became molecular: pharmacology, neurobiology, endocrinology, and even medicine, if several popularizers of the time are to be believed.[34]

This "molecularization" involved the movement of researchers. Some "leaders" of molecular biology had sharply reoriented their laboratories to the study of higher organisms, their development, or the functioning of their nervous systems. Other, younger molecular biologists—graduate or undergraduate students recently trained in the new discipline—were recruited to "classical" biology laboratories in order to provide new skills. This wave of young researchers trained in molecular biology arrived at the same time as new techniques from the discipline were introduced: the use of labeled molecules to follow RNA or protein synthesis, ultracentrifugation and chromatography to purify RNA or protein molecules, and electrophoresis to characterize them.

This introduction of researchers and techniques was a consequence of the invasion of various disciplines by the concepts and models of molecular biology. Whereas a traditional biologist would characterize a function in physiological or cellular terms, the molecular biologist—or the biologist recently retrained in the new discipline—would isolate the molecules (generally proteins) responsible for that function. In endocrinology (the science of hormones and their mechanism of action) or in neurobiology, the aim was henceforth to isolate the proteins onto which the hormone or the neurotransmitter bound (the receptors) and which were responsible for the observed cellular effect. Similarly, pharmacologists tried to isolate the macromolecules that were the target of drugs (these macromolecules were also called receptors). The notion of a "receptor" pre-dated molecular biology,[35] but it fitted perfectly into the models of the new science. Allosteric theory explained how the binding of a small molecule such as a hormone onto a receptor could alter the receptor's conformation and enable it to interact with other molecules and transport the hormonal or pharmacological signal to the inside of the cell. In medicine, the aim was to describe pathologies in molecular terms. The example had been given by Linus Pauling as early as 1949.[36] In an article published in *Science*, Pauling described the causal chain that linked the slight changes in the structure of the hemoglobin molecule observed in sickle cell anemia with the macroscopic changes in red blood cells that were responsible for the disease (see Chapter 12).

Nevertheless, this molecularization of biology should not be overestimated. Many laboratories—now sporting the term "molecular" in their title—carried on as before, with the same research objectives and using the same techniques. Jean-Paul Gaudillière has shown that, in France, the DGRST's "concerted action in favor of molecular biology" often had little effect on the research carried out by the laboratories involved.[37]

Furthermore, the molecularization of biology often involved the introduction of the techniques and models of biochemistry rather than those of molecular biology. Only a few groups began studying genes and gene regulation directly. In some cases, this can be explained by the persistence of research traditions and experimental systems that pre-dated molecular biology. In other cases, however, the absence of appropriate molecular techniques stopped researchers from turning to molecular studies or, when initial attempts were unsuccessful, led them to return to a biochemical approach.[38]

Such prudence was probably preferable to the "molecular" excesses that characterized some areas of research such as embryology, neurobiology, and immunology. All three disciplines almost simultaneously took up the idea that RNA might be responsible for intercellular information transfer.

In embryology, several groups argued that inductive effects—the possibility that cells of a given type could induce other cells to undergo differentiation—were produced by the transfer of RNA molecules from the inducing cells to the induced cells. Embryonic induction, which in the 1930s had led to a dead-end, thus received a new lease on life[39] (see Chapter 8).

In immunology, research from the beginning of the 1960s had shown that antibody synthesis was the result of the cooperation of different cell types: macrophages captured and transformed antigens, whereas antibodies were produced by lymphocytes. Several groups showed that the information necessary for the synthesis of specific antibodies was transferred from the macrophage to the lymphocyte as an antigen-RNA complex.[40]

But it was in neurobiology, and in particular in the study of memory, that this pseudo-molecular approach had its most striking successes, if only briefly.[41] At the beginning of the 1960s, studies on rats by H. Hyden in Sweden (using microspectrophotometric techniques) and by Louis and Josefa Flexner in the United States, and work on the small worm *Planaria* by James McConnell, showed that the synthesis of macromolecules (RNA or proteins) was required for memory formation, and in particular for the acquisition of conditioned reflexes.

In itself, this was hardly surprising. Memory, like any biological process, requires the transcription of a certain number of genes and the synthesis of the corresponding proteins. But the authors of these studies went much further, suggesting that there was a molecular coding of memories and behaviors in these animals. Both McConnell, working on *Planaria,* and Allan Jacobson, working on the rat, showed that it was possible to transfer *specific* behaviors from a trained animal to a naive one, in the form of molecules of RNA. These results, published by the most prestigious scientific journals and widely publicized in the media, provoked both excitement and major reservations. The very possibility that *Planaria* could learn[42] and the purity of the material used were questioned.

A number of researchers wrote a joint article in which they explained that they had been unable to reproduce the basic result—that is, the transfer of a conditioned behavior from a trained to an untrained rat by injecting RNA-enriched brain extracts.[43] At the end of the 1960s, however, George Ungar's results supported the possibility of such a transfer, but showed that the molecule responsible was not an RNA but an RNA-associated peptide. In the years that followed, other researchers also confirmed the earlier data, and Ungar generalized his initial observation to different kinds of acquired behavior—habituation to sounds, fear of the dark, and so on. He purified the peptides responsible and determined their amino acid sequence. The results suggested the existence of a mnemonic code. The peptides extracted from the brains of animals that had been conditioned to prefer a color (Ungar called these proteins "chromodiopsins") all had similar amino acid sequences, with a conserved part and a variable part.[44] According to Ungar, the conserved sequence corresponded to the preference behavior, whereas the variable sequence corresponded to the preferred color.

Ungar seemed to be on the threshold of discovering the code for memory. But the quality of his work was increasingly criticized, as was his behavior. He published less and less frequently in specialized journals, reserving his announcements for the press or for popular science magazines.[45]

These molecular theories of memory gradually disappeared. In 1975 the discovery of enkephalins and endorphins—endogenous analogues of morphine—some of which have an effect on memory, provided another interpretation of the memory-transfer data: Ungar's peptides belonged to the same family of molecules as the endorphins and had a nonspecific effect on memory.

In all three cases—the molecular bases of memory, embryonic induction, and the formation of antibodies—RNA molecules (or proteins in the case of Ungar's experiments) were seen as containers or transporters of information. They all integrated the new informational view of biology that characterized molecular biology. But the research was marked by a brutal reductionism, going directly from observed biological phenomena to macromolecules, ignoring what took place within the cell. In the experiments on antibody formation, as in the studies of embryonic induction, the observed effects implied that there was an interaction between cells. In the case of the acquisition of behavioral responses, it was known that different regions of the brain were involved in different behaviors, which therefore required the information to be transmitted through a neuronal network.

How can we explain this molecular "fashion"?[46] In the 1960s molecular biologists wanted to reduce as many biological phenomena as possible to the molecular level. As the years passed, however, the field of molecular biology became increasingly strong. Its struggle against the traditional biological disciplines became more subtle: molecular biologists infiltrated other disciplines, transformed them, and gradually took control. Once victory was attained, an armistice would be signed in which both sides agreed to respect the specificity of their respective approaches.

A complementary explanation that is more psychological and conceptual than sociological can also be proposed. Many molecular biologists who had received little or no training in the traditional biological disciplines turned with a certain naiveté and ignorance to the investigation of complex biological phenomena. The study of higher organisms led to the discovery that life in fact has a high degree of organization. This in turn led to the development of a new paradigm that was not molecular nor cellular nor physiological, but something different. To explain biological phenomena, this new approach made connections among the molecular level, the hierarchical structural organization of organisms, and the finalism of their behavior (see Chapter 21). This paradigm developed over several years and involved a few wrong turns, some of which have been described here.

CHAPTER 16

Genetic Engineering

Genetic engineering involves manipulating, isolating, characterizing, and modifying genes, as well as transferring them from one organism to another. It is not easy to describe the history of genetic engineering precisely, because it is a technology and not a science. The history of science often consists of a detailed account of a series of experiments and the theories that enabled them to be understood. But the development of a new technology cannot be described as a linear succession of elementary steps. When a new technology appears, a set of individually unimportant elements become linked in a network and take on a new significance. Some of these elements are clearly defined technological advances that were developed in order to fulfill their function in the network; others are old discoveries that had not previously been exploited, but that take on a new importance through their integration into the network. Finally, some elements in the network may be not discoveries per se but minor technical improvements that make a task possible by reducing the time and effort required. For example, techniques for sequencing nucleic acids existed before 1970, but they took so long and were so difficult that not until the development of rapid sequencing techniques between 1975 and 1977 did the field of genetic engineering begin to take off.

The exact frontiers of a technological network are also difficult to define: genetic engineering benefited from discoveries and new techniques in other areas of biology, immunology, and physiology (such as the discovery of monoclonal antibodies or the development of *in vitro* fertilization). These techniques were incorporated into the network of genetic engineering, enriched it, and modified it, while continuing to be a part of other technological networks.

The idea of deliberately changing the genetic make-up of an organism is as old as genetics itself. In 1927 this idea appeared to come a step closer when Hermann Muller, who had been Morgan's student, showed that it was possible to induce mutations by X-rays. This method was limited, however, because the mutations it induced were just as random as spontaneous mutations: their frequency could be increased, but not their specificity. According to Avery, only in pneumococcal transformation was it possible to induce "by chemical means . . . predictable and specific changes which thereafter could be transmitted in series as hereditary characters."[1] The later discovery of transformation in other bacteria, together with conjugation and transduction, all reinforced the idea that the genetic make-up of an organism could be deliberately modified.

The term "genetic engineering" goes back further than experiments on genetic manipulation. It originally referred to any controlled and deliberate modification of the genetic constitution of an organism, either by classical genetic experiments—crossing organisms carrying particular genes—or by future techniques that would involve the manipulation of isolated genes and their introduction into another organism without crossing.

Scientists soon realized that direct gene manipulation could have enormous practical applications. In 1958 Edward Tatum wrote:

> With a more complete understanding of the functioning and regulation of gene activity in development and differentiation, these processes may be more efficiently controlled and regulated, not only to exclude structural or metabolic errors in the developing organism but also to produce better organisms. . . . This might proceed in stages, from *in vitro* biosynthesis of better and more efficient enzymes to biosynthesis of the corresponding nucleic acid molecules, and to introduction of these molecules into the

> genome of organisms, either through injection, through introduction of viruses into germ cells, or through a process analogous to transformation. Alternatively, it may be possible to reach the same goal by a process involving directed mutation.[2]

In 1969 Joshua Lederberg wrote the genetics entry in the *Encyclopaedia Britannica Yearbook* and predicted the extraordinary applications of the new genetics: viruses would be used to introduce new genes into the chromosomes of humans or plants, thus providing a treatment for human genetic diseases or a way of improving crops.[3]

Viruses were widely seen as the vectors of choice for such an introduction of new genes: oncogenes were known to integrate themselves into the chromosomes of the infected cell and thus to transform its properties, like phage in bacteria (Chapter 15). But this power of transformation did not seem to be limited to viruses. Experiments on the direct transformation of the cells of higher organisms by foreign DNA had shown positive results as early as 1962, although they could not be replicated.[4] The isolation of purified fragments of DNA, and the total synthesis of a gene in 1970 (see Chapter 15), showed that the manipulation of genes was possible. Nothing, however, implied that molecular biology was on the threshold of a revolution even more thoroughgoing than that which had taken place twenty years earlier with the discovery of the double helix: the manipulation of genes still appeared to be a difficult objective, situated on a distant and ill-defined horizon.

Historians do not agree on the origins of genetic engineering. Some see it as a natural development, largely anticipated during the years that preceded the first experiments.[5] Others have argued that the discovery, characterization, and use of "restriction enzymes" that can cleave DNA at precise points played a decisive role. This seems to have been the opinion of the Nobel Committee, because in 1978 the first Nobel Prize for a discovery relating to genetic engineering was awarded to Werner Arber, Hamilton O. Smith, and Daniel Nathans. Arber had been the first to characterize restriction, that is, the ability of some bacteria to degrade—to "cleave"—any foreign DNA introduced into the bacterium, through the action of "restriction enzymes."[6] Smith had been one of the first to purify and characterize the properties of these enzymes,[7] whereas Nathans had used them to cleave DNA from the SV40 virus into fragments and thus make what would later be called a "physical map" of the viral genome, a prelude to the determination of the complete sequence of viral DNA.[8]

In fact, the experiment carried out at Stanford by David Jackson, Robert Symons, and Paul Berg and published in 1972 in the *Proceedings of the National Academy of Sciences* marked the beginning of genetic engineering.[9] In this article, Jackson, Symons, and Berg describe how they obtained *in vivo* a hybrid molecule containing both the DNA of the SV40 oncogene and the DNA of an altered form of phage λ (λ dvgal) that already included the *E. coli* galactose operon.

To carry out this study, Berg first cut the parental DNA molecules by using a restriction enzyme, *Eco*RI, provided by Herbert Boyer's group at the University of California at San Francisco. The ends of the λ dvgal fragment and of SV40 had been altered by the action of another enzyme—terminal transferase—which had added repeated sequences of adenine and thymine, respectively. These repeated sequences would pair together, leading to the association of the DNA from phage λ and the DNA from SV40. The addition of DNA polymerase and of an enzyme—ligase—that could close the circular molecules of DNA thus obtained terminated the operation and produced a circular molecule that was a cross between λ and SV40.

Strictly speaking, none of the steps used in Berg's experiment was original. Restriction enzymes had already been used to cleave SV40 DNA; the effects of the ligase and terminal transferase were already known—thanks, in particular, to the work of Arthur Kornberg's group on ligase (see Chapter 20).

The idea of fusing two molecules of DNA using repeated sequences of adenine and thymine via the action of terminal transferase had previously been proposed by Lederberg in a research project submitted to the National Institutes of Health.[10] Peter Lobban, a student in the same department as Berg, had begun trying to associate *in vitro* two DNA molecules from phage P22.[11] Berg's article was original, however, because it presented an important result, a wide range of techniques, and a clear outline of future perspectives. It described the first *in vitro* genetic recombination between DNA molecules from different species. Genetic recombination—the association of previously separate genes—is a natural process that occurs randomly in both bacteria and higher organisms. But it always takes place between genes carried by organisms of the same species, or, in the most extreme case, between the genes from a virus and the genes of the infected cell. In Berg's experiment, bacterial genes were linked to those from a monkey virus. Furthermore, Berg made *simultaneous* use of a large number of enzymes that were to form the basic tools of any experiment in genetic engineering: restriction enzymes to cleave the DNA and a ligase to close it up again, exonuclease to degrade it and DNA polymerase to repair it, while terminal transferase was used to make the ends of the DNA molecule "sticky."

Berg showed that the resultant DNA molecule could be integrated into the chromosomes of a mammalian cell. But because it also possessed DNA sequences from phage λ, it could replicate itself autonomously in bacteria. The hybrid molecule could thus be multiplied within these bacteria. In their article, Berg and his colleagues clearly described the dual use of genetic manipulation: the possibility of introducing genes into a given organism coupled with the use of bacteria to amplify hybrid molecules that had been created *in vitro.*

The article by Berg and his coworkers thus has the same founding value as Watson and Crick's 1953 article. In it the authors united a series of elements that might have been dispersed over a large number of experiments and articles. They grouped together a project, a series of techniques, and a result that was important in and of itself. The article was a scientific work of art.[12]

To give a complete picture of the first experiments in genetic engineering, three other results obtained around the same time need to be added:[13]

- In 1972, Janet Mertz and Ronald Davis discovered that the restriction enzyme *Eco*RI did not cleave DNA cleanly but left cohesive ends that could be stuck back together or associated with other DNA molecules that had been cleaved with the same restriction enzyme.[14] This result meant that it was possible to bypass terminal transferase, thus considerably simplifying *in vitro* recombination.

- The groups led by Herbert Boyer and Stanley Cohen used new bacterial vectors instead of phage λ. Plasmids—circular DNA molecules—had been discovered in 1965. They carry genes that confer resistance to antibiotics and can autonomously replicate in a bacterial cell, where they can often be found in hundreds of copies. The first recombinant plasmids, containing several resistance genes and/or DNA sequences from different species of bacteria, were made in 1973 and 1974. These plasmids could penetrate *E. coli* bacteria, where their genes could be expressed,[15] even if they came from other bacterial species such as *Salmonella* or *Staphylococcus.*[16]
- In 1974 an experiment carried out by Cohen, Boyer, and their coworkers showed that DNA from the toad *Xenopus laevis,* once introduced and copied in a bacterium, could be transcribed into RNA. The possibility of getting bacteria to synthesize a protein from a higher organism became more real.[17]

Many scientists were worried when Paul Berg announced his results prior to their publication in the *Proceedings of the National Academy of Sciences.* They feared that his research would lead to the development of *E. coli* bacteria carrying cancer-causing genes. These organisms might escape from the laboratory and spread their genes to the human population (which naturally carries *E. coli* bacteria). These fears were expressed publicly at a conference that took place in New Hampshire from June 11 to 16, 1973. Three months later, on September 21, 1973, Maxine Singer and Dieter Soll, who had chaired the New Hampshire conference, published a letter in *Science* in which they called for the creation of a committee to investigate the consequences of *in vitro* genetic manipulation.[18] In a letter sent to the *Proceedings of the National Academy of Sciences* and to *Science,* Paul Berg argued for at least a partial moratorium on such research and called for a conference that would define the conditions under which such experiments could be carried out.[19]

The conference took place from February 24 to 27, 1975, at Asilomar in

California and consisted primarily of a presentation of the experiments that had been carried out using the new technology. Only one day was devoted to their potential dangers. A report of these discussions, written by Paul Berg, David Baltimore, Sydney Brenner, Richard Roblin, and Maxine Singer, was published in *Science* on June 6, 1975.[20] The conference had tried to set out the risks associated with these new experimental techniques and the precautions that should be taken.

The conclusion was that work could begin again, but that physical and biological confinement measures would have to be adopted to limit the dangers. Physical confinement involved limiting all contact of the potential pathogen with the experimenter or with the environment. This could range from wearing a lab coat and gloves during the experiment, through obeying the sterility rules enforced in microbiology laboratories, up to the use of extractor hoods or even isolated rooms subject to lower barometric pressure than the regular environment and separated from it by an airlock. Biological confinement involved carrying out the *in vitro* recombination on organisms that had been modified so that they could not survive outside the laboratory.

Recombinant genetic experiments were classified according to whether they involved DNA extracted from bacteria, from animal viruses, or from eukaryotic cells. The main hazard was associated with the last kind of experiment: it was thought that the genomes of higher organisms contained inactive forms of potentially pathogenic viruses, in particular of tumor-causing viruses. Random manipulation of the DNA of higher organisms would run the risk of provoking the incorporation of these viral sequences into bacteria, and their subsequent spread to the human population. *In vitro* genetic recombination with animal viruses appeared to be less dangerous, except when the viruses were potentially carcinogenic. In this case, genetic manipulation would be authorized only if the sequences responsible for the cancerous transformation had already been removed.

Experiments on bacterial DNA were by far the least dangerous. Indeed, the exchange of genetic information between bacteria is a natural phenomenon. The only "risky" experiments were those involving DNA from highly pathogenic bacteria or DNA coding for toxin genes. These experiments were completely banned.

A hazard level was associated with each type of experiment, with the recommended confinement measures adapted to the degree of risk. According to the report's overall approach, the proposed measures were provisional and were expected to evolve as a function of increasing knowledge and improvements in biological confinement techniques.

The Asilomar recommendations were made into precise rules, set out for the United States by the NIH in January 1976. In Great Britain, the Williams report—published at the same time—set out a number of similar measures. In France, a DGRST commission set up in 1975 was charged with classifying projects and devising security rules. Given that there was no specifically French set of rules, the commission based its work on the recommendations made by the NIH and the Williams report.

A great media debate on recombinant DNA erupted, especially after the Asilomar conference, and reached its peak in 1976–1977.[21] There was in fact not one debate, but several, all of which were characterized by a substantial degree of confusion. The possibilities raised by the new experiments on *in vitro* genetic recombination, and the potential risks linked to these new techniques, changed over time, at least as they were expressed in the opposing positions.

It was soon apparent that the main fear of the scientific community in 1971–1972 was that it would be possible to introduce oncogenes into *E. coli.* Berg's experiment had in fact focused these fears because it had been carried out with DNA from an oncogenic virus. This fear was all the stronger because, between 1970 and 1975, most molecular biologists were influenced by viral models of cancer (see Chapter 19).

Although potentially dangerous, even tumor-causing, viruses had been studied for many years in a number of laboratories, such studies had never previously provoked any worries. With the first experiments in genetic engineering, a new group of molecular biologists, barely familiar with traditional biological disciplines or with the confinement measures used in laboratories that studied pathogenic organisms, began to study oncogenic viruses. The risk of an accident appeared much greater with these new experimenters, highly trained in the manipulation of DNA and restriction enzymes but largely ignorant of the precautionary measures traditionally used.

The fear that these genetic manipulations would lead to the appearance of new bacterial or animal species that could reproduce and perhaps develop and establish themselves, to the detriment of existing species, played only a minor role in these initial debates. Similarly, the fear that the manipulation of human genes would lead to a eugenic program aimed at the control and modification of the human genome was completely marginal.

After Asilomar, all that changed. The construction of new laboratories to carry out the most dangerous kinds of genetic experiments led to a series of confrontations. In the summer of 1976, when Harvard wanted to build such a laboratory, the mayor of Cambridge opposed the project, arguing that research should be carried out for the common good, not to get the Nobel Prize. After an enquiry that lasted several months, and after consulting experts, the town council overturned the mayor's decision and agreed to the construction of the laboratory.[22]

What had been a debate between experts exploded into the public domain. The arguments against recombinant DNA experiments became wide-ranging: the creation of new pathogenic species and the development of a lethal epidemic both appeared to be major dangers. In the longer term, recombinant DNA experiments ran the risk of encouraging mankind to control both itself and nature. It could also lead to the development of an exclusively genetic view of humanity and to, as a consequence, the development of eugenic policies.

The critics of recombinant DNA technology were part of a much larger current that was highly critical of science—as well as of its links with capitalism—and the "scientific-military-industrial complex." The fears that these critics expressed seemed to be justified by the interest shown in the new technology by the main chemical and pharmaceutical companies. For the opponents of genetic engineering, this interest both revealed the existing relations between science and business and heralded the development of even tighter links, which would tend to deprive scientific research of what little independence and creativity it still had.

Throughout this period, the scientific arguments of the supporters of recombinant DNA technology remained more or less constant. They claimed that the potential for beneficial applications justified fundamental research. They hoped that this technology would lead to the isolation and characterization of the genes of higher organisms and to the study of their organization and regulation. This new methodology was expected to show whether the mechanisms of cellular differentiation and intercellular communication—characteristic of higher organisms—involved new modes of genomic organization and regulation, or were merely minor variations of the mechanisms present in bacteria.[23]

The potential applications of the new techniques of genetic engineering proposed by its supporters were limited to a few examples that were regularly repeated throughout the long controversy. Genetic engineering would make it possible to

- reprogram bacterial cells so that they would synthesize clinically useful products from humans, such as insulin or interferon;
- modify plants to resist insects and diseases, or to fix nitrogen from the air (thus avoiding the use of expensive nitrogen-rich fertilizers)
- correct genetic "mistakes" in humans.

A growing number of scientists also hoped that research would be able to be carried out without hindrance or restrictive rules, which, by limiting the scope of research today, would also limit the number of practical applications tomorrow. In 1977 virtually the whole U.S. scientific community, led by the president of the Academy of Sciences, used these arguments as the basis of their opposition to research restrictions proposed by Senator Edward Kennedy of Massachusetts.

An unprecedented amount of experimental data was rapidly accumulated (see Chapter 18). The heat soon went out of the polemic, and security regulations, which originally had been extremely strict, were relaxed to previous levels. A number of more or less rational reasons were behind these changes.

The shift in attitude was not due to the development of more effective methods of biological confinement or to proof that the confinement methods employed were sufficient to prevent the dissemination of any pathogenic genes that might be isolated during recombinant DNA experiments. Rather, it was a direct result of several conferences that took place between 1976 and 1978 to estimate the risks of dissemination. Susan Wright has shown that the supporters of genetic engineering used these conferences to convince the scientific community and the general public that there was no danger.[24] The participants were limited to scientists who were favorable to recombinant DNA experimentation, and the discussion was centered solely on the risk of an epidemic due to recombinant bacteria and not on the potential hazards for technicians or researchers manipulating recombinant DNA. Furthermore, the reports of the conferences did not always reflect the debate. For example, they did not mention that a series of supplementary experiments had been commissioned or that the results were not yet available.

In fact, many of the discoveries made by genetic recombination tended to reduce the risk of such experiments (see Chapter 17). The risks associated with the transfer of a gene from a higher organism into bacteria were reduced when it was realized that the genes of higher organisms are split, interrupted by many long stretches of noncoding DNA, and that the excision of these sequences ("splicing") was necessary before a functional messenger RNA could be formed. Because bacteria lack the molecular machinery necessary for splicing, it would be impossible for the "introduced" gene to be expressed.

Despite these scientific arguments, the fact remained that experiments in genetic engineering could be hazardous and, in certain cases (which, though improbable, were not impossible), might lead to accidents affecting researchers and technicians in molecular biology laboratories, or even to an epidemic that might escape from the laboratory and affect the population. What gradually convinced people of the safety of genetic engineering was the lack of experimental accidents; in other words, people were convinced empirically but irrationally. This lack of accidents showed that the new technology did not *often* lead to problems with rapidly detectable consequences. Nothing proved, however, that rare but extremely serious accidents could not happen. As to the fear that genetic experimentation would lead to the modification of species in the future, to a clumsy attempt by humans to control their own evolution with the risk of a resurgence of eugenics, the early results of genetic engineering and the ongoing technical improvements would hardly change matters, even if the practical consequences of the results remained limited.

Early experiments had shown that it was possible to carry out genetic recombination *in vitro* and to introduce the product of this recombination into a bacterial cell. What was needed, however, before these experiments could open the road to the study of the genes of higher organisms, were techniques to isolate the genes.

Two complementary approaches were tried. The first, chosen by Tom Maniatis's group at Harvard,[25] involved going from proteins to genes, which made it necessary to choose a cell or an organ in which a particular protein was strongly synthesized. This was the case for the precursors of red blood cells, in which hemoglobin is abundantly synthesized. Messenger RNAs coding for hemoglobin are present at high levels in these cells, representing an important fraction of total mRNA. These molecules can be purified relatively simply (for example, by centrifugation), copied *in vitro* using reverse transcriptase (the enzyme discovered in retroviruses by Temin and Baltimore)—thus producing "complementary" DNA (cDNA)—and then inserted into a plasmid or phage and amplified in bacteria.[26]

This approach had two weaknesses. First, the genes obtained by this method were not functional because they did not carry regulatory signals (in particular, "promoter" DNA sequences) upstream from the gene, which enables the RNA polymerase to recognize the gene and to copy it into mRNA. Second, this procedure could only be used to isolate genes coding for abundant proteins.

Another technique, "cloning," was developed by David Hogness's group, which worked on *Drosophila* and wanted to isolate genes that classical genetic analyses had shown were important in the development of the fruit fly. The products of these genes, if they were known at all, were present in very small quantities. In this method, the whole genome is fragmented at random (cleaved by restriction enzymes), and the resultant fragments inserted into plasmids or phages, which are then incorporated into bacteria.[27] The result is a heterogeneous population of bacteria: each bacterium has a different phage plasmid containing a different part of the genome. This population is called a genomic "library." It is equally possible to create a cDNA library by copying all the mRNA molecules present in a cell into DNA, and then integrating these different cDNA molecules into plasmids, and then into bacteria.

Such libraries are useful only if those bacteria containing the plasmid or phage with the desired gene or stretch of cDNA can be isolated. The procedure is as follows: the bacteria that make up the library are separated by dilution and are then allowed to form isolated colonies on the surface of a petri dish. The colonies carrying the gene in question are then selected by molecular hybridization using a fragment of the DNA of the gene or of a similar gene. Thus, for example, in 1976 Maniatis's group isolated a cDNA and two years later was able to isolate the β-globin gene from genomic DNA libraries.[28] Molecular hybridization, developed by David Hogness, involves transferring the colonies onto a nitrocellulose filter, then denaturing them, that is, separating the two strands of plasmid DNA contained in the colonies.[29] These filters are then incubated with a fragment of DNA (cDNA in Maniatis's experiment) called a probe, which has previously been denatured and radioactively labeled. This fragment of DNA will pair up ("hybridize") with the plasmid DNA that contains a DNA sequence identical or analogous to that of the probe. The filters are then washed, and the presence of colonies carrying DNA fragments homologous to the probe is revealed by the trace they leave on a photographic film. The experimenter then simply has to choose the colonies that produced the positive result on the film, grow them in a test tube, and then isolate the recombinant plasmids. These plasmids, once cleaved by appropriate restriction enzymes, will liberate the desired DNA fragment.

The next step was the development of "expression" libraries. In such libraries the gene cloned in the phage (or plasmid) is expressed and its protein products are produced within the bacteria. This means that it is possible to use an antibody to detect the presence of this protein in bacterial colonies. The technique is virtually the same as that for molecular hybridization: the bacterial colonies are transferred from the petri dish onto a nitrocellulose filter that is then incubated with an antibody rather than with a radioactive probe.

By comparing libraries either from organisms that do or do not possess a particular gene, or from cells or organs that do or do not express the protein coded for by the gene, it is possible to isolate genomic DNA or cDNA even though no fragment of DNA from the gene has previously been isolated, nor any antibodies been raised against the corresponding protein.[30]

Genetic engineering made its first fundamental step forward with the *in vitro* recombination of DNA molecules. The second such step was cloning. But genetic engineering would not have developed so rapidly if new techniques, and the improvement of existing methods, had not simplified each step in the process. The following improvements—made by different research groups—had to be combined for genetic engineering to become an effective and efficient technology:

- More and more restriction enzymes were purified and characterized, each of which recognized a specific sequence and thus cleaved DNA into different size fragments. The commercialization of these enzymes, which rapidly followed their discovery, made research much easier. Mapping a DNA fragment—that is, locating the different restriction sites—became so easy that any student could do it. By 1978 approximately fifty restriction enzymes and their cleavage sites were already known.[31]
- The physical mapping of DNA molecules required a way of separating fragments following cleavage. Initial methods such as centrifugation were rapidly replaced by electrophoresis on polyacrylamide gels, then on agarose gels (agarose is a sugar polymer extracted from seaweed), with the DNA fragments being revealed by the addition of a fluorescent dye, ethidium bromide.[32]
- Once the fragments have been separated by electrophoresis, those that carry a given DNA sequence have to be isolated. This again involves molecular hybridization following transfer of the gel content to a nitrocellulose sheet.[33] A similar technique was developed to determine the size of RNA molecules corresponding to a given gene.[34]
- Increasingly effective "vectors"—phages or plasmids—were developed, containing several cleavage sites recognized by different restriction enzymes.

This meant that fragments of DNA could be cloned regardless of which restriction enzyme was used to cleave them. The original plasmids generally contained an antibiotic resistance gene, whereas more recent plasmids may contain several such genes, some of which can be inactivated by the insertion of the DNA fragment into the plasmid. As a result, bacteria that do not contain the plasmid will not develop in the presence of antibiotics; bacteria containing the plasmid alone will grow in the presence of any antibiotic, whereas those colonies containing the plasmid with the insert will grow only in the presence of certain antibiotics. This procedure makes it possible to isolate rapidly those bacteria that have integrated the recombinant plasmid, and to eliminate those colonies that contain only the original plasmid. The majority of plasmids used today derive from a prototype pBR322 constructed in 1977 by the Mexican scientist Francisco Bolivar.[35] Many plasmids, adapted to various objectives (cloning, expression of the genes they carry), were later created from this plasmid and sold by various companies.

- One of the limits of the first cloning experiments was that the plasmids or phages could incorporate only small DNA fragments (less than a few thousand bases). These vectors were not appropriate for cloning genes from higher organisms, which are generally tens of thousands of base pairs in length. "Cosmids"—vectors that are derived from plasmids but that could be "packaged" *in vitro* into the heads of bacteriophage—resolved this difficulty.[36] With the development of research on the human genome, new methods for cloning, mapping, and separating large DNA fragments were developed.

The first step in any research using genetic engineering is the isolation and amplification of the genes or purified DNA fragments. Once this stage has been completed, the DNA has to be characterized by making a physical map using restriction enzymes, and above all determining the order of its nucleotides, or its "sequence." This final operation became much easier following the simultaneous development in 1977 of two competing sequencing techniques, one by the Americans Allan Maxam and Walter Gilbert,[37] the other by the British scientist Frederick Sanger.[38]

Maxam and Gilbert's method used chemical reagents that cleave DNA at certain nucleotides, whereas Sanger used an enzyme-based method requiring DNA polymerase and specific inhibitors (see Chapter 20 for a fuller description of the development of Sanger's method). For both methods to be effective, polyacrylamide gel electrophoresis had to be improved. The development of thin gels and the use of new buffers substantially increased the resolution of electrophoresis, to the extent that DNA fragments differing by only one base could be distinguished. In a single electrophoresis it became possible to determine the sequence of DNA fragments more than 300 base pairs in length.

Genetic engineering also involves all those techniques that enable researchers to alter the sequence of isolated DNA fragments in order to modify either the protein they code for or the signals that regulate their expression. Various experiments using chemical mutagens were initially carried out, but it was only in 1978 that Michael Smith developed a method of directed mutagenesis using synthetic oligonucleotides (short fragments of single-strand DNA).[39] At the same time new, simpler, and quicker protocols for synthesizing oligonucleotides were developed,[40] and automatic synthesizer machines were soon commercialized. Genetic engineering put the oligonucleotides produced by these methods to a number of uses, including inserting new restriction sites in the vectors and adding regulatory signals.

If gene expression and function in higher organisms are to be studied, once the gene has been purified, characterized, and perhaps modified *in vitro*, it has to be reintegrated into the cells from which it was extracted. In the 1960s, experiments had shown that DNA could enter eukaryotic cells. Transfer was relatively inefficient, and the stability of the incorporated DNA was low. The use of animal viruses was not only potentially dangerous, as explained above, but it also limited the size of the DNA fragments that could be incorporated. In 1973 F. Graham and A. Van Der Erb developed a new technique that substantially improved the transfer of DNA into cells or, to use the molecular biologists' jargon, the effectiveness of "transfection."[41] Finally, in 1979, a method of cotransfection with a resistance gene was described, showing that it was possible to isolate cells that had integrated exogenous DNA[42] (this was analogous to the method developed several years earlier for bacteria). This experiment opened the road to the development of stable cell lines with a modified genetic constitution.

These techniques were the fundamentals of genetic engineering. In 1982, following a practical course taught at Cold Spring Harbor, a book called *Molecular Cloning: A Laboratory Manual*[43] was published that described all these techniques in extremely simple terms. The use of genetic engineering techniques became commonplace and marked the end of the developmental phase of the new technology.

The most important initial results using the new techniques were the discovery of the discontinuous structure of genes, the proof that genetic rearrangements exist in the cells of the immune system, and the demonstration that the genetic code is not, in fact, universal (see Chapter 17).

All these results threatened the "dogmas" of molecular biology. Some of these data were extremely important, whereas others related only to certain groups of species.[44] But what contributed most to the transformation of molecular biology was not these results but the simple practice of genetic engineering, the isolation of genes and the decoding of their sequences (see Chapter 18). This practice led to the characterization of the genes involved in the cancerous transformation of cells (see Chapter 19). This same technology enabled molecular biologists to isolate proteins that act as transcription factors and to characterize the master genes involved in the control of development.

In 1976 the first cDNA molecules, obtained by isolating mRNA and copying it into DNA by reverse transcriptase, were amplified following insertion in a plasmid, then isolated and characterized. As explained, these cDNA molecules corresponded to abundant RNA and proteins, such as β-globin,[45] insulin,[46] the rat growth hormone,[47] and human placental hormone.[48] These results made it possible to characterize the corresponding genes by screening genomic DNA libraries with cDNA probes. One of the first genes isolated in this way, in 1978, was the β-globin gene.[49]

The most sensational studies were not those that tried to isolate and characterize the genes of higher organisms, but those that sought to get bacteria to synthesize the products of these genes. These studies were carried out in an atmosphere of frantic competition between laboratories on the West and East coasts, between the University of California at San Francisco (and the biotechnology company Genentech) and Harvard University. The genes studied were those for which the protein might have important medical applications, such as insulin and interferon. Two different strategies were employed: either the DNA corresponding to the protein to be synthesized by the bacteria was synthesized *in vitro* (thanks to the genetic code, which allows the corresponding nucleotide triplet to be assigned to each amino acid), or cDNA was obtained after mRNA had been isolated and then copied into DNA by reverse transcriptase. In both cases, eukaryotic genes were linked to a series of bacterial regulatory sequences that were well known from the work of Jacob, Monod, and many others. Both approaches paid off, and between the end of 1977 and the beginning of 1979 a series of experiments confirmed that it was possible to express proteins from higher organisms in bacteria.

In the fall of 1977, Keiichi Itakura and Herbert Boyer of the University of California were able to get bacteria to synthesize human somatostatin,[50] a low–molecular weight hormone consisting of only thirteen amino acids. The corresponding gene had been artificially synthesized and then fused to a fragment of the lactose operon. The resultant product was a fusion protein, a hybrid of β-galactosidase and somatostatin. A later chemical reaction liberated the somatostatin.

This might seem of limited importance—a small protein was expressed as a hybrid molecule. But in fact this result showed that it was possible to express the protein of a higher organism in bacteria, thus paving the way for subsequent major developments.

This first success was quickly followed by the expression of human insulin using the same approach[51] (this development was itself preceded by the expression of insulin and of dihydrofolate reductase, an enzyme involved in nucleotide biosynthesis, from cloned cDNA).[52] Finally, in 1979 and 1980, bacterial synthesis of growth hormone[53] and then of biologically active interferon[54] reinforced the idea that the new technology could have medical implications.

To this list should also be added experiments that led to a genetic "movement" in the opposite direction, that is, to the insertion into higher organisms of genes that had been amplified in bacteria. As seen earlier, in 1973 a method for directly inserting DNA fragments into animal cells without using viruses had been developed and then improved by using selection genes (1979). In 1980 DNA was integrated into plant cells for the first time.[55] Because a plant can be reconstituted *in vitro* from a single cell, it thus became possible to obtain "transgenic" plants that incorporated new genetic information into their chromosomes. The creation of transgenic animals followed shortly afterward. In 1980 Frank Ruddle injected mouse embryos a few hours old with foreign DNA that then integrated into their chromosomes. After several rounds of cell division *in vitro,* the embryos were implanted into surrogate mothers, which, twenty days later, gave birth to a total of seventy-eight pups, two of which had integrated the foreign DNA into most of their cells.[56] Such "transgenic" mice were expected to be able to transmit the foreign DNA to their offspring, which was later shown to be the case. This transgenic animal experiment had been made possible by the development, a few years earlier, of methods for manipulating and growing embryos *in vitro.* These experiments ushered in the age of contemporary molecular biology.

Almost immediately, researchers began to apply this new technology to the early diagnosis of genetic disorders. By the 1960s, a certain number of genetic diseases or chromosomal disorders could be detected prenatally by culturing fetal cells present in the amniotic fluid and either looking for the products of the genes involved or observing the chromosomes—this was the principle of classic prenatal diagnosis. But if a given genetic disorder was not linked to a chromosome rearrangement that could be detected under the microscope, or affected a gene that was expressed not in the cells to be found in the amniotic fluid but only in specialized fetal tissues, prenatal diagnosis was impossible.

With the new molecular technology, this difficulty disappeared. In 1976 Yuet Wai Kan of the University of California at San Francisco isolated DNA from normal subjects and from patients affected by a particularly serious form of anemia resulting from the loss of genes coding for α-globin.[57] He hybridized this DNA with a labeled cDNA corresponding to α-globin. By measuring the quantity of hybridized labeled cDNA it was possible to distinguish normal subjects from patients with the disorder. Kan repeated the experiment with DNA extracted from amniotic cells and was able to show that the fetus concerned was not affected, which was confirmed at birth. A completely reliable diagnosis had been made, without any blood being taken from the fetus: all the cells of an individual carry genes for α-globin, whether it is expressed or not. In the years that followed, researchers greatly improved this technique and applied it to many other genetic disorders.

CHAPTER 17

Split Genes and Splicing

The existence of the genetic code implied an alignment, a perfect "colinearity" between the genetic message and the primary protein structure: this had been experimentally confirmed in bacteria, with all the precision that genetics could provide.[1] Despite the fact that higher organisms are much more complex than bacteria, most molecular biologists thought that in all organisms the same steps enabled the information contained in genes to be expressed as proteins. The discovery that genes are interrupted in cells with a nucleus, and that the coding DNA sequences are separated by noncoding sequences, took place in different stages.

During the Cold Spring Harbor Symposium of 1977, several groups—in particular those of Richard Roberts and Phillip Sharp—reported that in the adenovirus (a respiratory tract virus), the definitive mRNA was a copy of noncontiguous DNA sequences; it was made up of a long fragment that coded for viral proteins, with, at one end, a series of small fragments that had been copied from distant parts of the DNA sequence.[2] By using restriction enzymes, gel electrophoresis separation of the DNA fragments, and hybridization, scientists were able to draw up a detailed map. Electron microscopy allowed them to visualize directly the hybridization complexes formed between RNA and DNA.

Following this initial discovery, it appeared that the fragmentation of the genetic message was limited to the extremity of the mRNA that did not code for proteins, but that contains the leader sequence that initiates the translation of RNA into proteins. Molecular biologists thought that this result would be limited to viruses. The need to compress genetic information into an extremely small volume had led viruses to adopt unusual solutions: in the φX174 bacteriophage it had recently been shown that the same DNA fragment could code for several completely different proteins by reading the nucleotides in different phases.[3]

The result obtained with adenovirus was confirmed by the analysis of another small animal virus, SV40,[4] then extended to the genes of higher organisms.[5] By hybridizing mRNA to genomic DNA, several groups showed that in the genes of higher organisms, even protein-coding DNA sequences were interrupted by noncoding regions: to use the nomenclature introduced by the American scientist Walter Gilbert, the genes of higher organisms look like mosaics, consisting of a series of "exons" (DNA sequences that can be subsequently found in the mRNA) and "introns" (silent DNA sequences that are absent from the final mRNA and have no apparent function).[6] This result was confirmed for all subsequently studied genes, such as hemoglobin, immunoglobin, ovalbumin, and collagen.[7] The number of introns varied but was always large, and could be as high as sixty for some genes, whereas the "intronic sequences" could represent up to 95 percent of the gene. The fragmentation of genes into introns and exons was not the exception; it was the rule. Apart from yeast, in which both the number and the size of introns seemed to be limited, all cells with a nucleus had mosaic genes. Even mitochondria—the "power stations" of nucleate cells—which have their own DNA molecule coding for ribosomal and transfer RNA and a few specific proteins, have "split" genes.[8]

Using radioactive isotopes, molecular biologists showed that the gene was first copied into a long RNA molecule. This RNA, which was very short-lived, was then converted into mRNA.[9] During this process, the RNA was "spliced": the RNA sequences corresponding to the introns were eliminated, while the RNA fragments corresponding to the exons were lined up. This process of cleaving and ligation of RNA fragments took place in the cell nucleus: the only RNA molecules to be found in the cytoplasm were the definitive mRNA molecules.

This discovery of "split" genes hit the world of molecular biology like a bombshell; it was a "minirevolution."[10] As seen earlier, however, a large number of indexes had already suggested that the mechanisms that led from DNA to proteins were much more complex in higher organisms than in bacteria. For example, it was already known that mRNA molecules were modified at the 3′ (right-hand) extremity.[11] Molecular biologists had already begun to characterize the RNA molecules present in the nucleus—HnRNA—and shown that they were much larger than mRNA molecules. The work of G. P. Georgiev's group, followed by a number of different experiments, had led to the hypothesis that mRNA molecules corresponded to the 3′ part of HnRNA and were thus produced by the breakdown of the HnRNA molecule.[12]

The fact that these scientists did not discover the fragmentation of genes can perhaps be explained by the difficulty of their experiments, which were carried out before the development of genetic engineering. The lack of a mechanism explaining how fragments of DNA that were far apart in the genome could be close together in the same mRNA molecule perhaps led some experimenters to ignore—or perhaps exclude—some observations. The French biologist Pierre Chambon realized that prior to the 1977 Cold Spring Harbor meeting he already had experimental data suggesting that the ovalbumin gene was split, but they had not been exploited.[13]

The true explanation appeared too absurd even to be considered. Indeed, the existence of interrupted genes was completely unexpected. Studies of bacteria had suggested that the genetic message was clear, well punctuated, and economically written. In fact, it turned out to be long and complex. Nevertheless, molecular biologists were enthusiastic about the discovery of splicing: they had finally discovered a clear difference between bacteria and higher organisms. From a neo-Darwinian point of view, the evolutionary conservation or invention of a process as aberrant as gene splitting could be explained only if the process played an essential role in the cell, for example, by regulating gene expression.

Several models were proposed along these lines. Control of splicing could stop the expression of genes that had been transcribed by mistake. In this model, splicing played the role of a final check,[14] but in most models the process was given a far more important regulatory role. Some researchers transposed the regulatory mechanisms of transcription that had been discovered in bacteria: specific proteins—inhibiting or activating—were thought to fix onto the HnRNA and to modulate splicing. Other models were proposed that involved the simultaneous splicing of several different genes and could explain the coordinated changes of expression observed during cell differentiation and in embryogenesis: a new "path" of differentiation was the result of a new splicing enzyme.[15] These models adopted the spirit, if not the letter, of Roy Britten and Eric Davidson's model.[16]

Even more daring hypotheses were put forward: while studying introns contained in yeast mitochondrial DNA, the French geneticist Piotr Slonimski (who had been Boris Ephrussi's student) noticed that if splicing did not take place normally, a hybrid protein could be formed from the genetic information in the intron, together with information from the neighboring exon. Given that genetic experiments had shown that changes to the intron could alter splicing, Slonimski suggested that this hybrid protein, which he called maturase or m-protein, participated in the splicing of its own mRNA. Slonimski's model thus stated that a gene contained the information required for its own splicing. This model allowed for spontaneous autoregulation because the maturase, by acting and by splicing the HnRNA, eliminated the RNA that produced it.[17]

Slonimski adapted this model—first developed for mitochondrial genes—to nuclear genes[18] and both enriched and extended it: the study of the amino acid composition of "theoretical nuclear maturases" suggested that these proteins were to be found at the nuclear membrane and thus coordinated the splicing of HnRNA with the passage of RNA through the pores of the nucleus. If there were only a limited number of pores, the different RNA maturases might have to compete for a place near one of the pores. This competition would be won by those maturases that had been synthesized first. Slonimski's model suggested that cellular regulation might depend on the initial state of the cell. If this cell was an egg—the embryo at the beginning of its development—this would explain the existence of nongenetic hereditary constraints that affect development but are not the result of the action of embryonic genes. An analogous model was used by another French molecular biologist, Antoine Danchin, to explain cancerous transformation and ageing in cells.[19]

Thus in the short space of time between 1977 and 1980, and on the basis of a proven fact—the existence of split genes—molecular biologists developed a series of models that gave this phenomenon a fundamental regulatory role. Rather than simply revealing the imagination of experimenters, these models expressed their frustrations at the long absence of major discoveries that could explain the mechanisms of cell differentiation and embryonic development.

Unfortunately, experimental results did not confirm the hopes of the molecular biologists. None of the proposed models turned out to be false, but their biological significance is in fact extremely limited: Slonimski's model, for example, applies only to a few mitochondrial introns. The regulation of splicing does indeed take place during certain steps of cell differentiation, but this is a qualitative regulation (the cell controls production of two different mRNAs from a single genomic fragment) rather than a quantitative regulation (the quantitative level of expression remains essentially that of the transcription of DNA into RNA). This regulation of splicing plays only a secondary, minor role in controlling gene expression. Wherever it has been observed, it corresponds to a principle of economy that enables the cell to make two slightly different proteins from a single gene.

Molecular biologists abandoned the idea that an explanation of the complexity of higher organisms could be found in the fact that their genes were "split." Molecular biology was thus "complicated" but not "transformed" by the discovery of exons and introns. It was the biochemical approach to splicing that produced the biggest surprises: a series of different mechanisms was discovered, requiring a complex molecular machinery and implying that the three-dimensional structure of RNA played an important role.

In some ribosomal RNAs, splicing turned out to be spontaneous, requiring no external protein, being catalyzed simply by the RNA molecule.[20] In itself, this result might seem unimportant, because it was limited to the splicing of a few types of RNA. In fact, it was extremely interesting: together with data on ribonuclease P,[21] it showed that nucleic acids could catalyze a chemical reaction, thus behaving like an enzyme.

Up until this point, molecular biology had strictly separated the functions of biological macromolecules: nucleic acids were involved in stocking and transferring genetic information, whereas proteins had structural and catalytic functions. Unfortunately, this division made it impossible to understand the origin of life: not only would it mean that the two types of macromolecules must have appeared at the same time, but, right from the outset, there must have been a correspondence between the information carried in the nucleic acid molecules and the polypeptide chains that make up proteins; in other words, there must have been a functional genetic code.[22]

With the discovery of the catalytic role of RNA, it became possible to imagine that, at the beginning of life, the same molecule carried genetic information and had a catalytic function. The current world must have been preceded by an "RNA world," and the division of labor between the two main classes of biological macromolecules must have taken place later in evolution.[23]

The discovery of interrupted genes obliged molecular biologists to turn to the study of the mechanisms of evolution.[24] It was easy to explain the origin of point mutations where one amino acid is replaced by another—Watson and Crick had proposed a model in their 1953 articles (see Chapter 11)—but it was much more difficult to explain the insertion of amino acids into a protein. The existence of introns could account for this: a point mutation at the junction of an exon and an intron that effectively removed the distinction between the two sequences would add amino acids to the middle of the protein.[25]

A number of observations confirmed that, over the course of evolution, there had been insertions or deletions of amino acids at intron-exon splice junctions. Furthermore, in most cases these junctions appeared to correspond to those parts of the protein chain that were on the outside of the macromolecule.[26] Changes to these junctions—which would not change the overall structure of the protein—could lead to the creation of new catalytic sites.[27]

Differential splicing, mentioned earlier, permitted the evolution of a new form of protein without leading to the disappearance of the old form.[28] Researchers thus discovered a mechanism that created new forms without altering previous function. S. Ohno had ascribed a similar function to the duplication of genes.[29] Through gene duplication, the existence of interrupted genes had also favored the appearance of genes that coded for complex proteins. For the product of such duplications to be functional, the DNA separating the copies has to be precisely removed. Splicing carries out this excision quite "naturally."[30]

Some molecular biologists hypothesized that an analogous evolutionary process could have associated fragments of genes—exons coding for protein substructures. The complex proteins currently found would thus be the result of the assembly of several different elementary protein domains by recombination of the exons coding for them.[31] Evolution could use this extremely powerful "erector set" to "tinker" with great effect.[32]

Once again, however, seductive hypotheses were only partly verified. There are many examples of gene duplication in the course of evolution. Some complex proteins, such as immunoglobulins, are the result of the repeated duplication of a single exon. But though the recombination of exons can explain the appearance of some proteins, such as those forming the extracellular matrix of the organism, it is by no means a general rule. Many exons do not correspond to precise structural domains;[33] the position of introns may differ in two structurally similar proteins;[34] and similar structural domains may have different evolutionary origins.

The theory of "exon shuffling" does not seem to be as general as was first hoped: evolution has not always used the same bricks to build complex protein structures; in fact it has probably reinvented the same basic elements several times over. The appearance of introns certainly made the recombination and duplication of genes easier, but it did not profoundly change the speed of evolution.

Molecular biologists initially thought that introns and exons were specific to cells having a nucleus. The existence of introns in mitochondrial genes, probably of bacterial origin, suggested that this was not the case. The exon-shuffling model was in close agreement with the idea that introns must have existed in all organisms during the first phases of evolution. The role of exon shuffling was probably more fundamental at these early stages than later during evolution. From this point of view, today's bacteria could be considered the most highly evolved organisms, having eliminated all the "junk DNA" to be found in introns.[35]

It is particularly striking that the discovery of split genes was rapidly integrated into evolutionary models. It was a paradox: many molecular biologists present themselves as the most ardent supporters of the neo-Darwinian theory of evolution, yet to suggest new mechanisms implies that this theory, at least in its canonical form, cannot explain the evolution of higher organisms. Furthermore, the new mechanisms, such as exon shuffling, are not faithful to the spirit of Darwinism, because they provide no immediate advantage to the organisms that possess them. The maintenance of introns can only be a neutral process, merely tolerated by evolution.[36]

In the late 1970s and 1980s, four more discoveries questioned the "dogmas" of classical molecular biology. In 1977 Claude Jacq, working in the laboratory of G. G. Brownlee at the MRC in Cambridge, England, discovered the existence of "pseudogenes" that were structurally similar to genes but were apparently silent. They were initially considered evolutionary residues, the inactive product of gene duplication. Three years later, however, Jacq and his colleagues discovered that some pseudogenes had lost their introns, as though they had been formed from mRNA by the action of reverse transcriptase.[37] Ten years after the development of Howard Temin's proto-oncogene model (see Chapter 15, p. 170 ff.), there was a new threat that the genome could integrate information from the cytoplasm, raising with it the specter of Lamarckism.

The menace gradually evaporated, just as it had ten years earlier. Spliced pseudogenes did indeed exist, but they turned out to be of limited importance, perhaps the result of the parasitic action of a retroviral reverse transcriptase.

Another surprise came from the study of the mitochondrial genome. The comparison of the DNA and protein sequences showed that the meaning of several "words" in the genetic code was not the same in mitochondria as in *E. coli* and nuclear genomes.[38] For example, the UGA codon normally stops protein synthesis, but in mitochondria it leads to the incorporation of the amino acid tryptophan. This "deviation" from the universal genetic code is, however, restricted to a few codons, to some mitochondria, and to the nuclear genome of some unicellular organisms such as *Tetrahymena*[39] and paramecia.[40]

Other unicellular organisms—trypanosomes—revealed the existence of a new mechanism called "editing,"[41] which alters the transfer of information between DNA and proteins. RNA molecules, produced by the faithful transcription of the gene, are not functional. They lack some nucleotides, generally uridines, which are added after transcription. The number of missing nucleotides varies between one and several hundred.

There are several different versions of editing, some of which can be observed in higher organisms. But the phenomenon is most important in the mitochondria of single-cell organisms, where it has been most closely studied. Editing is linked to splicing—small RNA molecules ("guides") act as nucleotide donors.[42] Editing again confirmed the catalytic properties of RNA, but, despite its strangeness, its "quantitative" importance remained limited.

The final discovery solved, at last, the mystery of antibody synthesis. As noted, Linus Pauling thought he had solved this problem by hypothesizing that the antigen played an active role in antibody synthesis. But this model had not been supported experimentally, and above all had suffered because it did not fit with the "dogmas" of molecular biology. Francis Crick, in his famous 1957 lecture "On Protein Synthesis," still held that the synthesis of both antibodies and adaptive enzymes was an exception to a general rule, a case where the action of genes was insufficient to "form" proteins.[43] Following Jacob and Monod's work, the synthesis of adaptive enzymes was seen to agree with the dominant view of molecular biology. The formation of antibodies ought also to comply with the new conceptual framework. To this end, in 1959 Joshua Lederberg proposed the following hypotheses in an article in *Science:*[44]

- The structure of antibodies depends solely on their amino acid sequence.
- This sequence, which is different for different antibodies, is genetically determined.
- Because there cannot be as many genes as there are antibody molecules, the genes responsible for antibody synthesis must have mutated, and mutated differently, during the formation of the cells that produce antibodies.

Lederberg's theory can be understood only in the framework of the clonal theory proposed shortly before by the Australian scientist Frank MacFarlane Burnet.[45] Burnet was not only an immunologist; after training as a virologist he had become interested in the biology of cancer. He was convinced that both tumor development and viral epidemics were due to the growth of one or more clones, that is, to the existence of a few cells or viruses that could multiply extremely rapidly. Burnet transposed these ideas to antibody synthesis: each antibody was synthesized by one or more cellular clones that multiplied when they encountered an antigen with an affinity for the antibody they synthesized. These clones were "preprogrammed" to make the appropriate antibody. To explain how the cell could start to multiply in the presence of the antigen, Burnet suggested that the cell might carry its antibody on its surface. Contact with the antigen would modify the cell membrane and start cell replication. Antibodies cannot be made against the components of the organism, because the corresponding clones are eliminated during development.

Burnet's theory was soon accepted by the scientific community, and in particular by the "molecular biologists." But it needs to be emphasized that Pauling's theory was not abandoned because of any experimental data: the rare sequences of antibody molecules known in 1959 were not sufficient to reject the hypothesis that all antibodies were formed from the same polypeptide in differently folded forms. A year earlier Lederberg had carried out an experiment that suggested that each cell in the immune system made antibodies that could recognize only one kind of antigen.[46] The fact that this was an isolated result, together with the technical difficulties involved in carrying out the experiment, meant that its impact was limited. Lederberg himself recognized that Pauling's theory had become unacceptable, not in its detail, but rather in its spirit. The new theory was "more closely analogous to current conceptions of genically controlled specificity of other proteins."[47] The theory had to be selective and not instructive, faithful to the neo-Darwinian spirit of molecular biology.[48]

In fact, Burnet and Lederberg had merely pushed the problem of diversity back to genes and DNA. Their proposed solution—the existence of somatic mutations in immunoglobulin genes—remained unproved until 1976, when it was corrected following the direct analysis of the genes responsible for making antibodies. The role of somatic mutations was maintained, but was associated with a highly complex process of the rearrangement of gene fragments: the study of antibody molecules had shown that they contained two polypeptide chains, each of which was formed of variable and constant regions. In 1976 Nobumichi Hozumi and Susumu Tonegawa showed that DNA sequences coding for the variable and constant regions were far apart in embryonic cells, but close together in antibody-producing cells: the genes coding for antibody molecules had been rearranged during development.[49]

Subsequent studies showed the complexity of these rearrangements, which involve several different genetic segments. This genomic rearrangement during development turned out to be limited to the cells of the immune system, taking place in genes coding for immunoglobulins in B lymphocytes and in genes coding for receptors in T lymphocytes.[50] Jacob and Monod's view that the genome was stable during development (see Chapter 14) was thus not seriously altered. The movement of fragments of DNA in the genome is of limited importance and does not play a major role either in regulation or in the evolution of higher organisms. But these rearrangements do play a fundamental role in monocellular organisms such as ciliates.

CHAPTER 18

A New Molecular Biology

The new molecular biology has become a way of "reading" life.[1] "Classical" molecular biology had shown the importance of the genetic code, of the information linked to nucleotide sequences. Sanger, followed by many others, sequenced proteins and substituted the linear sequence of their amino acids for their structural complexity. But this "classical" molecular biology was centered on proteins. In experimental terms, studies of the structure, function, and specificity of proteins came before studies of their amino acid sequence, anticipating the study of the nucleotide sequence in the genes that coded for these proteins.

On the surface, none of this changed with the introduction of genetic engineering techniques. Proteins and their three-dimensional structure were still considered the specific agents of biological processes, acting in cells as enzymes, as structural proteins, or as regulatory proteins. For the molecular biologists of yesterday and today a protein is sufficiently understood only when its three-dimensional structure has been determined by X-ray diffraction.

But in experimental practice, this is no longer the case. The number of genes for which the sequence is known is growing exponentially. From these sequences, it is possible to deduce the amino acid sequence of the proteins that are potentially coded by these genes. By contrast, the number of proteins for which the three-dimensional structure is known is growing very slowly. The molecular biologist has had to make the most of the one-dimensional information that is directly available, and be satisfied with a linear reading of the book of life.

What can be done with such a text, written in a decipherable but unknown language, in which the three-dimensional meaning of the words remains hidden? Like an archaeologist who knows how to read a language without understanding its meaning, the molecular biologist looks for regularities, repetitions, and resemblances. This approach has led to the identification of families of proteins with similar structures, making it possible to guess the function of proteins that are known only through their corresponding gene. The description of such similarities thus has a heuristic value. Furthermore, it raises the question of the "ancestry" of different proteins and, quite naturally, leads to the study of evolution.

The fact that these similarities frequently concern only one part of the protein implies that protein structure should be seen as composite, formed of several independent modules. A similar modular conception has arisen following the study of DNA sequences that are found upstream of the gene and are necessary for the regulation of the gene's activity. Many such sequences have been described; some of them, situated far away from the gene, are called "enhancer" sequences. Each represents a binding site of one or more protein factors.

Repeated observations have shown that the regulation of gene expression in eukaryotic cells is due to the action of a combination of proteins that bind upstream of the gene. Jacob and Monod's models have had to be re-elaborated: regulation is now seen as the result not of the action of a few highly specific inhibitor proteins, but of the combination of a large number of low-specificity activating proteins. Understanding this combinatory process will undoubtedly require the development of new conceptual tools.

Through its discoveries, molecular biology has metamorphosed from a science of observation into a science of intervention and action. The book of life, even before it has been understood from a structural point of view, can be altered. The effects of molecular modifications can be analyzed at two levels.

At the molecular level the experimenter can induce subtle changes by using directed mutagenesis, replacing one amino acid with another. If the protein under study is well known and its three-dimensional structure has been determined, these substitutions will make it possible to test different models of the function of the protein. If the protein is an enzyme, it will be possible to study its catalytic mechanism. If it is a regulatory protein, one can determine which amino acids interact with DNA. The comparison of what a molecular biologist could do at the beginning of the 1970s and what is possible today gives a measure of how much things have developed. In the past, it was possible to imagine a functional scenario, following the determination of the structure. But the model could be tested only indirectly, and remained largely hypothetical. Today's molecular biologists can not only test models experimentally but also quantify the relative contribution of each amino acid or each bond to catalysis or binding. These interventionist possibilities have led, almost automatically, to the creation of proteins with novel catalytic properties.

At the level of the organism, the molecular biologist can also look for the effect of induced molecular modifications by introducing the modified form of the gene into an egg or cell, thus creating a transgenic organism. This makes it possible to determine *in vivo* the function of each protein or DNA fragment. It is equally possible to inactivate a gene or to alter its product by homologous recombination (the replacement of the normal copy by an altered copy) or to modify its level of expression—and all this in whichever organism is most appropriate to the subject. With genetic engineering, biology became an experimental science at the molecular level. Even if the models remain relatively simple and over-mechanistic, the same distance separates the old and the new molecular biology as separates any science of observation from a science of action.

One way of highlighting the progress that has been made is to return to the problem raised by Niels Bohr sixty years ago, in the conference "Light and Life."[2] Bohr pointed out that any molecular study of an organism required the destruction of its structures and the death of the organism and the cells that contain the molecules. Because of this problem, even the most efficient *in vitro* systems can always be criticized for not completely reproducing the properties of the living organism.

With the development of genetic engineering, this problem—the inherent limitation of any molecular study of biology—has disappeared. The role of molecules that form an organism can be studied without killing the organism. This was an essential development, but one that came late: few people any longer doubted that the molecular study of biology would be sufficient to understand the elementary principles of the functioning of life.

Linguistic metaphors have often been used to describe the concepts and theories of molecular biology. Molecular biology became integrated into the new informational vision of the world that became dominant at the end of the 1940s. These metaphors took on their full meaning in the context of genetic engineering. To decipher the language of life, molecular biologists use the same method that a child uses to learn to speak. They imitate the sentences of life, modify them, and see how organisms react to these new sentences. Through this dialogue, scientists learn the terms that are permitted and those that are not—the syntax of life. Sequencing the human genome and the genomes of other organisms will open the "book of life." Will it, however, enable us to understand the language of life?

CHAPTER 19

The Discovery of Oncogenes

Between 1975 and 1985, a series of experiments showed that cancer, whatever its direct cause, is the result of the activation, by modification or over-expression, of a highly conserved family of genes called oncogenes.[1] These genes are involved in the control of cell division; their products form part of the regulatory network that relays signals from outside the cells to the nucleus and enables cells to adapt their rate of division to the needs of the organism. These discoveries form what can be called "the oncogene paradigm." The slow and complex development of this paradigm was possible only because of the tools of genetic engineering. The paradigm shows how molecular biology became integrated into other biological disciplines to form a "new biology."

Many molecular biologists turned to the study of small tumor-causing animal viruses in the hopes of finding a model system for studying animal cells and a tool for understanding the molecular bases of cancerous transformation. Among these viruses, the most widely studied were the RNA viruses, called type C retroviruses because of their appearance under the electron microscope. The RNA that forms them is recopied into a double-strand DNA molecule through the action of reverse transcriptase, and this DNA integrates itself into the genome in an inactive form called a provirus. Some of these viruses cause tumors, others do not.

Around 1975, the dominant model of the origin of cancer was that proposed by two American scientists, Robert Huebner and George Todaro. Their model gave retroviruses a key role in the process. Huebner and Todaro argued that such viruses existed in *all* genomes: silent and nonexpressed, they were nevertheless transmitted from generation to generation. Under the effect of mutagens, or by the action of other DNA or RNA viruses, these silent proviruses could be activated and become oncogenic. According to this model, all cancers, be they spontaneous, induced by chemical agents, or induced by viruses, are the result of the activation of proviruses.[2]

Another model, called the protovirus model, had been proposed by Howard Temin, the discoverer of reverse transcriptase. This model was not very different from that of Huebner and Todaro, and also implicated proviruses in the development of cancers. But for Temin, proviruses (which he called protoviruses) were not silent tumor-inducing structures that passively waited in the genome to be activated, but a part of the normal cellular genome, involved in the transfer of information from DNA to RNA and vice versa[3] (see Chapter 15).

In 1973 Edward Scolnick showed that a sarcoma-inducing tumor virus could be produced by the recombination of genetic material from various sources. He suggested that the development of a type C transforming virus could involve the capture by these viruses of oncogenic information provided by the cell.[4]

These different theories about the origin of cancer were largely based on data from bacterial genetics obtained ten years earlier. In 1960 Jacob had suggested that studies of lysogeny and bacterial regulation mechanisms might provide models for understanding cancer.[5] Huebner and Todaro's provirus model was the adaptation of the prophage concept to tumor viruses. The possibility that a virus could take over and "transduce" genes near its chromosomal insertion site had been proved for bacteriophage,[6] and Scolnick merely extended the idea to tumor viruses. Furthermore, it had been shown that the integration of a prophage in the genome of a bacterium could disturb the functioning of neighboring genes. Several research projects on tumor viruses tried to characterize the genomic virus-insertion sites, on the grounds that cancer might also be the result of the alteration of the functioning of normal genes in the cell following the integration of a tumor virus.

These models were not the only ones to try to explain the development of cancer. In a lecture given in 1974, Temin reviewed five different models that explained the origin of cancer, including the simplest, according to which a cancer-causing gene would appear by the simple mutation of a normal cellular gene.[7] Surprisingly enough, these different models were not in competition, and were not even mutually exclusive.

Thus the very idea of cellular oncogenes—normal genes that can become transforming genes responsible for cancer following mutation or transduction by a virus—had already been proposed and even accepted by the biological community prior to the discovery of the first cellular oncogene, "sarc," which was soon renamed *src*. The experiments that led to this discovery were described in two articles published in the *Journal of Molecular Biology* (1975) and *Nature* (1976).[8]

The experimental system used was the virus that causes Rous sarcoma—the first tumor virus to have been discovered, in 1911 in the chicken. In the early 1970s, studies had shown that, in this virus, a single gene was responsible for the malignant transformation of infected cells. The French scientist Dominique Stehelin and the Americans Harold Varmus and Michael Bishop isolated a cDNA probe corresponding to the viral gene responsible for cancerous transformation. The hybridization techniques used for the subsequent steps in the study were not original, having already been used by a number of other groups with the same objective. Stehelin and his colleagues showed that normal chicken cells carried copies of a gene that was very similar to the transforming gene of the Rous sarcoma virus. Using differential hybridization, they compared the structure of this gene in different birds and showed that it had varied over the course of evolution in a way similar to other avian genes.

In their conclusion, the authors suggested that the cellular gene they had discovered must play a role in the regulation of cell growth and in embryonic development. This important experiment was well received by other scientists. Nevertheless, the result was not considered revolutionary. Even its authors interpreted it in the context of the models of cancer discussed above. If Huebner and Todaro's model was accepted, the *src* cellular gene might come from a type C provirus in which the other viral sequences had evolved to the extent that they could no longer hybridize with the rest of the Rous sarcoma virus. The result could also be interpreted in the light of results previously obtained by Scolnick: the *src* gene was a normal cellular gene that had been "transduced" by the Rous sarcoma virus.

Stehelin and his colleagues did not generalize these results to other vertebrate species: they only claimed that birds carried sequences similar to the *src* transforming gene. In particular, no similar sequence was found in mammals. The fact that this negative result was at first readily accepted may seem surprising: in today's oncogene paradigm the products of oncogenes participate in the control of cell division in all cells, and because of this ubiquitous function have been highly conserved in the course of evolution. It was two years later, in September 1978, that Deborah Spector and her colleagues showed that there was a gene homologous to *src* in mammals and in fish.[9] Subsequent publications[10] and historical descriptions of the discovery of oncogenes[11] do not indicate that a two-year delay was necessary before the oncogene discovered by Stehelin and his colleagues was accepted as a ubiquitous component of the cells of higher organisms.

The importance of the discovery of the *src* oncogene was also hidden by the disappointing results of the study of the expression of this cellular oncogene: it was expressed at a low but identical level in all avian cells, be they normal or cancerous, and its expression did not change over the course of development.[12] These two characteristics were not those that were expected of a transforming, tumor-inducing gene. As shown in Chapter 14, cancer was considered to be a process of deregulation:[13] the expected result was thus that *src* would be strongly expressed in tumor cells. Furthermore, malignant transformation is often accompanied by the re-expression of genes that are active during development. Indeed, tumor cells, like embryonic cells, are able to multiply extremely rapidly. Researchers therefore expected that oncogenes should also be strongly expressed in the embryo. The properties of the *src* gene thus appeared very strange.

In the years that followed, these results for *src* were extended to other retrovirus transforming genes. But this generalization was insufficient to lead to the birth of a "paradigm" of cellular oncogenes. The new paradigm was established between 1981 and 1984 and was the result of a set of simultaneous discoveries showing that the same genes were involved in "spontaneous" cancers and in those induced by chemical agents or by viruses; and that oncogene-coded proteins were involved in normal cell growth.

A new experimental approach developed by the American scientists Robert Weinberg and Geoffrey Cooper, known as the "transfection assay," played an important part in the establishment of this new paradigm. These experiments were the natural consequence of the work carried out on RNA and DNA tumor viruses. They were initially developed to answer the following question: was it possible, after the integration of a virus into the genome as a provirus, to recover the virus (by isolation and fragmentation of genomic DNA) in an infectious form? The DNA of cells that had been transformed by a virus were cleaved by restriction enzymes and the resultant fragments were added to normal, untransformed cells in the presence of calcium, which enabled the fragments to be incorporated into the cells and integrated into the genome. Many colonies of transformed cells appeared among "transfected" cells, confirming that the "provirus" had conserved its transforming properties.

Researchers now wondered if, using the same approach, normal cells could be transformed with DNA extracted from cells that had been transformed by chemical agents, or with the DNA from "spontaneous" human tumors.[14] Previous studies had shown that known tumor-inducing chemical agents were also, after modification within the organism, mutagenic.[15] This suggested that the transformation of cells by chemical agents was the result of mutations induced in one or probably several normal genes. In the long debate between supporters and opponents of the genetic origin of cancer, these experiments tipped the balance in favor of the former.

Cooper and Weinberg's results were unambiguous:[16] in a significant fraction of chemically transformed cells or tumors, it was possible to isolate DNA fragments that were able to transform normal cells (fibroblasts) after transfection. But DNA extracted from normal cells was nontransforming.[17] Thanks to the new tools of genetic engineering, it was not long before these transforming DNA fragments were isolated (cloned) and characterized (in particular sequenced). Unexpectedly, these transforming sequences turned out to be identical to a number of cellular genes that had previously been characterized by their homology with genes from transforming retroviruses (the *ras* oncogenes).[18] Less than a year later, scientists characterized the mutation that led to transformation in DNA extracted from a bladder cancer. This mutation, which had vital consequences for cell physiology, was the result of a minor alteration in the coding part of a *ras* gene: a change in a single nucleotide led to the replacement of the twelfth amino acid in the protein by another amino acid.[19]

Other transfection experiments revealed modifications in different cellular oncogenes. Two results were particularly unexpected: the same genes were found to be involved in very different forms of cancer, and it was discovered that an oncogene mutation may not involve the promoter—and thus not change gene expression—but may instead alter its structure. It was soon shown that during tumor formation, the same oncogenes could be activated by three other molecular mechanisms, including chromosomal translocation.[20]

The simultaneous proof that, in different tumors, using different mechanisms, the same small group of genes was activated and became transforming showed that these genes were key actors in this transformation. But on its own this result would not have been enough to give rise to the oncogene "paradigm"; it was merely a first step in the establishment of this paradigm. The second step involved a series of discoveries about oncogene function that explained their central role in transformation and cancerogenesis. These discoveries were largely a matter of chance:

- In 1983 PDGF, a growth factor (a small protein that can stimulate cell division *in vitro)* present in blood platelets was sequenced. Computer comparison of this sequence with known protein or nucleotide sequences showed that it was similar to that of a transforming gene present in the simian sarcoma virus.[21] This suggested that the viral oncogene was a slightly altered version of one of the normal genes coding for PDGF. A small change in a growth factor or its expression was thus enough to induce cellular transformation and cancer.

- In 1984 an oncogene, *erb-B,* carried by a retrovirus that causes avian erythroblastosis, was sequenced and was found to be virtually identical to part of the sequence of the receptor protein for another growth factor, EGF.[22] This link between oncogenes and growth-factor receptors was not completely unexpected: in 1980 several groups had shown that some growth-factor receptors and oncogenes had the same enzymatic activity and induced similar modifications in a number of cellular proteins.
- In 1984 the sequence of a G protein was determined for the first time. G proteins play an essential role in the transmission of information from extracellular receptors to the inside of the cell. They turned out to be similar to products of *ras* oncogenes.[23]
- Other oncogenes, such as *myc* and *fos,* also play a role in the regulatory pathways that control cell division: adding growth factors induces the transitory expression of these oncogenes and leads to cell division.[24]

The development of this functional aspect of the oncogene paradigm was supported by the discovery that the structure of cellular oncogenes had been highly conserved during evolution: in 1983 genes that were partially related—both in functional and in structural terms—to *ras* genes were discovered in yeast.[25] From a neo-Darwinian point of view (shared by most molecular biologists) this result was entirely expected for cellular components playing a key role in growth and in cell division.

At the same time, scientists took a decisive step forward in their understanding of the regulatory networks controlling cell division. In addition to the "second messenger" that was already known, cyclic AMP (Chapter 15, note 32; a second messenger is a small molecule that "relays" an external regulatory signal inside the cell), new second messengers were discovered and the mechanisms leading to their production were determined. The characterization of an enzyme, protein kinase C, which is activated by these second messengers and intervenes in signal transduction, took place at the same time as the discovery of the role of cellular oncogenes in these regulatory pathways.[26]

The third step that led to the acceptance of the oncogene paradigm—anticipated by François Cuzin (France) and Robert Kamen (Britain),[27] then brilliantly realized by Robert Weinberg and his coworkers—was the proof that cancer developed through the cooperation of different oncogenes.[28] The discovery that a series of mutations in different oncogenes was required before a cancer could develop helped physicians to accept the new paradigm. It was widely held in medical circles that cancer could not be the result of a simple mutation: the frequency of point mutations was much higher than the frequency of cancer. Furthermore, the fact that cancer rates increase with age, and that the disease takes a long time to develop (for example, after a strong dose of radiation, the development of cancer takes several years), suggested that several steps were necessary for tumor formation.

The existence of cooperation between oncogenes led to their provisional classification into two groups—immortalizing and transforming oncogenes.[29] The activation of the former—such as *myc* or *fos*—led cultured cells to divide indefinitely. The products of these genes act in the nucleus and directly control gene transcription and cell division. The transformation of the normal form of the oncogene into an altered, transforming version is the result of the over-expression of the oncogene. Together with immortalizing oncogenes, transforming oncogenes such as *ras* are responsible for cellular transformation, for the acquisition of cancerous properties. They code for products that are present in the cell membrane or cytoplasm. A structural modification of these proteins is required for the cells to go from a normal to a transformed state.

Many facts have been omitted from this brief history of the oncogene paradigm. For example, recessive oncogenes (also known as anti-oncogenes or tumor-suppressor genes), first described in 1971,[30] have not been mentioned. Transformation occurs only when both copies of the gene are altered (inactivated). The first member of this new family of tumor-causing genes to be characterized, in 1986,[31] was the retinoblastoma gene. The discovery of tumor-suppressors suggested how DNA tumor viruses might function: the products of these viruses inhibited the action of tumor-suppressors. Nevertheless, their discovery did not lead to the development of a new paradigm. The products of tumor-suppressors, like those of oncogenes, are involved in the same regulatory networks that control cell division.

As noted, the oncogene paradigm developed relatively late, between 1981 and 1984. This was the result of two sets of simultaneous and chance discoveries: the same genes were found to be involved in different kinds of cancer, and the products of these oncogenes were shown to lead to uncontrolled cell division, to the kind of proliferation that characterizes cancer. These discoveries provided a rational explanation of the development of cancer.

In accepting the oncogene paradigm, most biologists agreed that studying this small family of genes was the best way to understand cancerous transformation. Alterations in these genes and their functioning were considered to be the cause of cancer. The oncogene paradigm is thus both a community of objects (cellular oncogenes and their products) and a community of methods that permits the characterization of the structure and function of these oncogenes and their products.

How can the virtually unanimous acceptance of this paradigm be explained, given that, for more than a century, the field had seen so many controversies? Several practical reasons explain its acceptance. The new paradigm made it possible, at least theoretically, to develop a series of diagnostic and prognostic tests. It opened an important area of biomedical research to molecular biologists. It enabled virologists to use the skills they had acquired in their long and often fruitless search for cancer viruses.[32]

There were also two more fundamental reasons. The first was that many of the discoveries that led to the establishment of the paradigm were in fact completely unexpected. Computer-assisted sequence comparison played a key heuristic role in the development of the paradigm's functional aspects. Such chance discoveries reinforce the realist philosophy of scientists (and in particular of biologists) and make them confident of the data. This heuristic value of sequence comparisons is characteristic of the new form of molecular biology that was ushered in by genetic engineering (see Chapter 18).

To understand the second reason the paradigm was accepted so easily, one must remember that the word "cancer" does not mean the same thing in different biological disciplines. The observations and facts that a new theory would have to explain were not the same for different specialists. For physicians, a theory of cancer would have to explain its progressive development and its multifactorial origin; it would also have to include the well-established correlation between the mutagenic and the oncogenic effects of a variety of chemical compounds or physical treatments. For biochemists, a model of cancer would have to explain the many biochemical and structural changes that take place in cancerous cells. For cell biologists, cancer was a disease that disturbed intercellular communication. Furthermore, many cancers appeared to be linked to the existence of chromosomal break-points, followed by translocation between chromosomes of the resultant fragments—something that could be seen under the microscope.[33] For virologists, most human cancers probably had a viral origin. Even if this were not the case, the mechanisms of nonviral oncogenesis had to be identical to those of virally induced cancers. Finally, for molecular biologists, the isolation and characterization of the genes involved in cancer were a natural objective.

Interestingly, but probably not surprisingly,[34] the oncogene paradigm led to the unification of these different conceptions of cancer, through the alliance of the concepts of molecular biology with those of other biological disciplines. The changes that take place in tumor cells are indeed biochemical, but they are linked to changes in a small number of genes. Very few cancers are caused by viruses, but viruses change cells by the same mechanisms—the alteration of oncogenes—as radiation or chemical mutagens. Cancerous changes directly affect cells and not the whole organism, but the cells are altered in their ability to communicate with the external milieu and with other cells. Several oncogenes have to be modified to obtain cell transformation. Chromosomal translocations are directly linked to a change in the expression of oncogenes.

Even opposing visions of cancer were reconciled in this new paradigm: regulatory and structural changes in genes are required for full cell transformation; changes affecting whole chromosomes or isolated genes can lead to either the activation or the modification of cellular oncogenes. Most biologists thought that the site of tumor-virus integration into the genome played a key role in transformation (as previously explained, this idea had its origin in studies of bacterial genetics from the 1960s, and contradicted results that showed that some tumor viruses themselves had transforming genes). These two different visions were united in the new paradigm. The activation of cellular oncogenes could be the result of their incorporation (and, perhaps, modification) within a retrovirus, or the integration, close to these genes, of a retrovirus that had just activated their expression.

It is striking that in the ten years that separated the discovery of the *src* oncogene from the final acceptance of the paradigm, no one system, no particular gene, played a dominant role. The discovery of *src* was, of course, extremely important, but for a long time its precise mechanism of action remained unknown. Similarly, the discovery of changes in the *ras* oncogene in human tumors had a substantial psychological impact, but it took nearly ten years for scientists to begin to understand the role of *ras* in cancerous transformation. To this day, the precise mechanism of action of *myc* and the effects of *myc* amplification on cell division remain poorly understood.

The American virologist Peter Duesberg repeatedly underlined these problems.[35] But the absence of clear answers did not stop researchers from making progress. Some areas of research continued to advance, others reached a dead-end and were replaced by new subjects. Some questions are still unanswered, and whole areas of research have been abandoned. Unresolved problems and unexplained data are buried in the back of researchers' minds, waiting to be incorporated into new models. Scientific research progresses anarchically, like a macrophage or a phagocyte: it extends its pseudopods when it can, but looks for another way when it meets an obstacle. Despite this chaotic approach, scientific knowledge, like the phagocyte, manages to go forward.

The development of the oncogene paradigm also reveals two other, highly specific aspects of modern biological research:

- The technology available to the biologist is very sophisticated: isolating a gene and characterizing its modifications have become routine. The heuristic value of the simple comparison of DNA sequences has been added to this technological power. It took only a few years for the main oncogenes (or anti-oncogenes) to be characterized and their function inferred.
- The history of oncogenes also underlines the limits of molecular knowledge. The greater the number of oncogenes characterized and the better understood their modifications and functions, the less biologists are able to distinguish the fundamental from the secondary, and to put some order into the ever-expanding amount of data. The simplicity of the 1980s paradigm has been diluted in the complexity of the subsequent accumulation of data. This is analogous to the difficulty encountered, for example, in the study of development. It raises two important questions: how can we go beyond today's molecular systematics, and how can we develop a new logic of life?

CHAPTER 20

From DNA Polymerase to the Amplification of DNA

In 1983 Kary B. Mullis developed a technique for amplifying DNA called the polymerase chain reaction (PCR).[1] PCR can amplify virtually any DNA fragment, even if it is present in only trace amounts in a biological sample, thus allowing it to be characterized. It can aid forensic medicine by characterizing DNA molecules present in biological samples such as hair, traces of blood, and so on. It is sufficiently sensitive to permit the detection and characterization of the rare DNA molecules that persist in animal or human remains thousands of years old. This technique also makes possible a genetic diagnosis on the basis of a single cell, such as, for example, a cell of a very young embryo prior to implantation in the uterus. Finally, it permits the early detection of bacterial or viral infections.[2]

Despite the fact that PCR appeared several years after the main techniques of genetic engineering, it shares several essential characteristics with them. Although it does not, in and of itself, open new experimental horizons—it merely "amplifies" DNA, which can also be done by other, more time-consuming methods such as cloning—in practice, it has made possible a number of experiments that were previously impossible.

PCR is a "postmature" development (see Chapter 2, p. 21); there is no reason it should not have been developed at any point after the early 1960s. The principle of PCR is based on the properties of DNA polymerase, which was isolated and characterized in 1955 by Arthur Kornberg.

The history of PCR leads back to the path that opened the way to the development of molecular biology. In particular, it involves a series of studies that provided biochemical confirmation of the double helix model of DNA proposed by Watson and Crick in 1953. It highlights once again the difficult problem of the relationship of molecular biology and biochemistry. It also emphasizes the important degree of technological continuity between the molecular biological studies of the 1940s and the work of the 1980s.

When scientists discovered that an enzyme could copy molecules of DNA *in vitro,* they removed one of the weaknesses of Watson and Crick's 1953 model—it was based entirely on the remarkable properties of enzymes, which, up until that point, had not actually been shown to exist. Furthermore, the discovery of DNA polymerase also confirmed a number of characteristics of the model. It is only briefly described by the historians of molecular biology: H. F. Judson gives it only a few lines (a transcription of Crick's reaction on hearing Kornberg's initial results at a 1956 conference on the chemical bases of heredity, at John Hopkins);[3] and in *A Century of DNA,* Franklin Portugal and Jack Cohen provide a brief and prudent description.[4]

This discretion contrasts sharply with the official recognition of Kornberg's work. In 1959, together with Severo Ochoa, he was awarded the Nobel Prize for his work on enzymes that form polynucleotides. The Nobel Prize Committee was remarkable both for its rapidity and for its perspicacity. Kornberg's work on the purification and characterization of DNA polymerase began only in 1955; the first results were presented in 1956, and the first articles were published in 1958.

Kornberg's scientific career was characteristic of a physician who decided to turn to research during the Second World War.[5] His first studies were on nutrition and vitamins. In 1945, supervised by S. Ochoa, Kornberg helped purify two enzymes that are essential for the cell's energy metabolism. After a short stay in Getty and Carl Cori's laboratory at Saint Louis—

called "the Mecca of enzymology"[6] because the laboratory studied very complex enzymes involved in the synthesis of sugar polymers—Kornberg returned to Washington, where he studied an enzyme able to synthesize one of the essential coenzymes to be found in the cell, NAD (a dinucleotide). This work made him familiar with nucleotide metabolism. From 1953 to 1955, he characterized several enzymes involved in nucleotide biosynthesis.

From Kornberg's autobiographical writings it is difficult to understand what led him, in 1954, to begin studying the enzymes that make DNA and RNA. Was it a natural prolongation of his previous studies? Did he want to follow the same path as other eminent biochemists, such as the Coris, who, after working on the enzymes involved in sugar metabolism, then studied enzymes that polymerize sugars into long polysaccharide chains? Or, despite Kornberg's protestations to the contrary,[7] was he influenced by molecular biology and the impact of Watson and Crick's 1953 model?

In 1954 Kornberg began characterizing and purifying two enzymes, one involved in RNA synthesis, the other—able to make DNA— by using a radioactively labeled nucleotide, thymidine. As the source of the enzymes Kornberg chose an extract of *E. coli* bacteria because, as he said, "the rapidly growing *Escherichia coli* cell had become a favored object of biochemical and genetic studies."[8] Initial results showed that a small but reproducible fraction of the thymidine was incorporated into a polymer that was probably DNA.

These promising initial studies were interrupted by the announcement that Severo Ochoa and Marianne Grunberg-Manago had discovered an enzyme—polynucleotide phosphorylase—that could polymerize long chains of RNA in the absence of any matrix.[9] Kornberg, together with Uriel Littauer, purified this enzyme from *E. coli* bacteria.[10] This enzyme later turned out to be extremely useful for synthesizing RNA molecules with different nucleotide compositions, used in experiments aimed at deciphering the genetic code (Chapter 12). But Kornberg and his colleagues soon realized that this enzyme incorporated nucleotides randomly, and thus could not play any role in intracellular information transfer. They returned to their work on DNA polymerase and succeeded in purifying the enzyme several thousand times. Using this purified enzyme, they were able to show that all four deoxyribonucleotides were necessary for *in vitro* DNA synthesis. The DNA matrix was also indispensable, and all synthesis stopped once an enzyme that digested DNA, DNase, was added. By contrast, adding an enzyme that digested RNA, RNase, had no effect, showing that the replication of DNA did not require a passage through an RNA intermediary.

With even purer enzyme fractions, Kornberg was able to replicate DNA twentyfold *in vitro:* at the end of synthesis, the parental DNA represented only 5 percent of the DNA present in the tube.[11]

He found that the best substrate for replication was a simple strand of DNA, which agreed with the replication model initially proposed by Watson and Crick (see Chapter 11). Furthermore, in the newly synthesized DNA, the percentage of A was always equal to that of T, and the percentage of G was always equal to that of C. The base composition of the newly synthesized DNA was identical to that of the DNA in the test tube at the beginning of the experiment.[12]

All these results showed that the DNA present at the beginning served not only as a primer for DNA synthesis (like enzymes that need a primer made of a few links of sugar to make polysaccharides), but also as a matrix that was faithfully copied by DNA polymerase.

A sophisticated labeling experiment with radioactive nucleotides, followed by digestion with highly specific nucleases, made it possible to determine the nature of the nucleotides situated close to the incorporated radioactive nucleotides. This experiment confirmed that the DNA synthesized by DNA polymerase was identical to the matrix DNA, but further showed that the newly synthesized strand was oriented in the opposite direction from the strand that had served as the matrix,[13] a result that again agreed with Watson and Crick's model.

Kornberg's experiments were remarkably elegant. They also represented a considerable amount of work: in order to have the necessary molecular tools, Kornberg had been obliged to purify several enzymes involved in nucleotide metabolism. Above all, these experiments showed that the extraordinary replicatory ability of DNA was based on the properties of a single enzyme, DNA polymerase, able to choose with a high level of accuracy among the nucleotides present in the environment to pair up the bases, as suggested by Watson and Crick.

Nothing in Kornberg's work proved that the DNA synthesized *in vitro* had the same biological properties as the parental DNA. The quick decision of the Nobel Committee to award him a prize was undoubtedly a result of the biochemists' determination to affirm their participation in the final stages of the race for the "secret" of life that was represented by the development of molecular biology. If the Nobel Prize was awarded for Kornberg's confirmation of Watson and Crick's model, it would have been more logical to give the prize to them first, and then to Kornberg! In fact, Watson and Crick received the Nobel Prize only in 1962, whereas Hershey, Luria, and Delbrück had to wait until 1969 before the importance of the phage group was officially recognized.

The fact that Severo Ochoa was a co-winner of the Nobel Prize for his discovery of polynucleotide phosphorylase—an enzyme whose precise physiological role remains poorly understood but which is probably involved in the breakdown of RNA—shows that above all the gesture was intended to reward those whose successful research had shown that intracellular transformations of informational molecules were also part of enzymology, and thus of biochemistry.

Instead of opposing molecular biology and biochemistry, these disciplines should be seen as having complementary roles: the molecular biologist deciphered the main pathways of information transfer, whereas the biochemist dealt with the "details" of the molecular machinery. These "details" required enzymes that carried out astonishing—and somewhat magical—tasks.[14] This division of labor provided molecular biologists with a confirmation and a "concretization" of the mechanisms that they had hypothesized; for the biochemists, it confirmed that everything in the cell was biochemical, from intermediate metabolism to gene replication.

Nevertheless, Kornberg's studies lacked a final experimental confirmation: they needed to show that DNA synthesized *in vitro* was indeed "active." But the "activity" of DNA, the proof of its informational role, was not easy to detect *in vitro*. The best experimental system for proving this was transformation (see Chapter 3): it should be possible to amplify the transforming factor using Kornberg's purified enzyme.

When Avery's results were presented earlier, particular emphasis was given to the difficulties involved in controlling and quantifying transformation experiments. The experiments aimed at amplifying transforming DNA by the action of DNA polymerase gave negative results.[15]

It was only in 1967 that Kornberg was able to multiply *in vitro* the DNA of a phage, φX174.[16] This experiment worked only because of the addition of another enzyme, DNA ligase, which closed the newly synthesized molecules. This experiment, which the scientists involved modestly described as the twenty-third and twenty-fourth contributions to the enzymatic synthesis of DNA, was presented by journalists as the creation of life in a test tube, and as the most important scientific discovery of 1967.[17] This gap between the reactions of journalists (and the public) and the attitude of scientists shows the change that had taken place in the minds of biologists. The definition of life had been altered: for biologists, replicating a virus *in vitro* is not the equivalent of creating life. Life does not lie in molecules; it is to be found in the complexity of the systems under study.

This change is all the more striking if the reactions of scientists and the public in 1967 are compared with the response to Stanley's crystallization of the tobacco mosaic virus thirty years earlier: at that time, the general feeling was that life had been reduced to the properties of the molecules that make up organisms.

This late but enthusiastic reception for Kornberg's work explains the disappointment that followed the discovery, in 1969, that his purified enzyme was not involved in DNA replication in *E. coli*.[18] The media circus that followed,[19] and lasted more than two years, can be explained—like the enthusiastic welcome for the replication of phage φX174 *in vitro*—by the long period in which molecular biology was "crossing the desert" (1965–1972). Although molecular biology had precise concepts, its tools were still inadequate. The fact that few discoveries were made led to an over-interpretation of the rare data, and to a continual questioning of the dogmas of the young science under the slightest experimental pretext. The debate that accompanied the discovery of reverse transcriptase was a good example of this situation.

Later experiments confirmed that Kornberg's purified enzyme was not essential for DNA replication, but instead intervened in DNA repair. But the enzymes involved in *in vivo* DNA replication have a structure that is close to Kornberg's DNA polymerase, and, most important, they function according to the same principles. The discovery that Kornberg's DNA polymerase had only a minor physiological role did not call into question the fact that enzymes are able to replicate DNA, as set out in Watson and Crick's model.

The polymerization activity of Kornberg's enzyme was used to incorporate labeled nucleotides into an unlabeled molecule and thus to make radioactive probes that would permit the detection of homologous RNA or DNA molecules by hybridization on gels or in libraries. Kornberg's enzyme is all the more useful because, in addition to its polymerizing activity, it possesses a degradation activity that enables it to eliminate mismatched nucleotides *in vivo,* and *in vitro* to replace nonradioactive nucleotides with their labeled equivalents.[20]

There is also another use for Kornberg's DNA polymerase: in 1977 Frederick Sanger showed that it could be used to determine the sequence of DNA molecules. (PCR is an offshoot of this sequencing technique.) Like many discoveries, this was the fruit of a meeting between an idea and pure chance.[21] Sanger wanted to follow, as polymerase proceeds along the DNA, the incorporation of nucleotides into the newly synthesized DNA molecule. Chance played a part in the preparation of the experiment. Because Sanger and his colleagues wanted to incorporate as much radioactivity as possible into the synthesized DNA, they used three unlabeled nucleotides, while the fourth—"X"—was labeled, but was added at a very low ("limiting") concentration. They found that this low concentration sometimes made the enzyme stop and detach where it should have incorporated an X nucleotide. The researchers could separate strands of different lengths using electrophoresis on acrylamide gels, and directly determine the relative position of the different X nucleotides. The experiment was then carried out with the three other nucleotides to derive the complete sequence. The principle of the sequencing technique had been found: stop the elongation of the DNA strand at a given nucleotide and then determine the length of the different fragments thus obtained.

Sanger subsequently made a number of improvements to this method. Its "canonical" form is as follows:[22] the DNA fragment to be sequenced is cloned into an M13 phage, which can exist as a single strand of DNA. Recombinant single-strand phages are isolated, and an oligonucleotide complementary to the phage DNA, and which hybridizes close to the insertion site of the fragment to be sequenced, is added. This oligonucleotide is simply a primer for Kornberg's DNA polymerase, which cannot start synthesizing DNA *ex nihilo,* but can only elongate previously existing strands. The four nucleotides are then added, together with a small quantity of an analogue of one of the nucleotides—dideoxynucleotide—which, when it is randomly inserted into the sequence in place of the proper nucleotide, immediately halts DNA synthesis. The experiment is carried out with four different analogues, each corresponding to one of the nucleotides, and the synthesized strands are analyzed by electrophoresis.

There is another method for sequencing DNA, named after its inventors, Maxam and Gilbert.[23] This method uses chemical reagents to cleave DNA at precise positions. Both methods are equally simple, but Sanger's method was subsequently adopted for automatic DNA sequencing.

This example again underlines the important role that enzymes play in molecular biological research and as tools in genetic engineering. Researchers use enzymes because of their remarkable specificity: because DNA polymerase faithfully copies DNA strands, it can be used to determine the sequence. But previous studies by Kornberg's group had also shown that this enzyme could be "tricked" into incorporating analogues, such as dideoxynucleotides, into the DNA strand, in place of normal nucleotides. A precise understanding of DNA polymerase was a key prerequisite for its use.

Kary Mullis has provided a detailed account of the discovery of PCR.[24] The idea came to him one Friday evening in April 1983, when he was driving on the hilly roads of California, on the way to the chalet where he was to spend the weekend.

Mullis had received a doctorate in biochemistry, and in 1979 had been recruited by the Cetus biotechnology company to make oligonucleotides for use as probes (see Chapter 16). The arrival on the market of oligonucleotide synthesizers had freed researchers like Mullis for other projects. While he was driving that evening, he was thinking of how to develop a technique that, on the basis of an extremely small biological sample, would be able to determine the identity of a nucleotide at a precise position in the DNA molecule. The project was interesting because many genetic diseases are produced by the substitution of a single nucleotide at a precise position in a gene. Such a technique would mean that it would be possible to make a genetic diagnosis on the basis of a very small biological sample.

Mullis's idea was to use Sanger's sequencing technique: the first step would be to synthesize an oligonucleotide immediately next to the nucleotide that was to be determined. The two strands of DNA would be separated by heating, and the oligonucleotide would be hybridized with the complementary strand. This oligonucleotide would function as a primer for the DNA polymerase. The four radioactively labeled dideoxynucleotides would then be added one after the other, but the only one that would be incorporated would correspond to the one at the position of interest.

The idea was a good one, but it suffered from the fact that oligonucleotide fixation is not always specific. Mullis decided that it would be possible to confirm the result by synthesizing a second oligonucleotide corresponding to the sequence immediately downstream of the nucleotide, on the complementary DNA strand (the two strands of a DNA molecule are oriented in opposite directions). This oligonucleotide would hybridize with the DNA strand, and would be elongated by incorporation of a dideoxynucleotide that was complementary to the dideoxynucleotide incorporated in the first experiment. The two experiments should thus lead to complementary results.

There was one further difficulty: the DNA sample might contain free nucleotides that could be incorporated instead of the dideoxynucleotide. Mullis's idea was to do the experiment in two stages: first, he would add the oligonucleotides, but not the dideoxynucleotides. The DNA polymerase would use the nucleotides present in the milieu to elongate the oligonucleotides. Once the reaction was complete (and the free nucleotides used up), all that would be necessary would be to heat the mixture to separate the two strands of DNA from the oligonucleotides of varying lengths to which they had hybridized, and then rehybridize the DNA with new oligonucleotides, this time adding the dideoxynucleotides.

But what would happen if the oligonucleotides had been sufficiently elongated to hybridize with the second oligonucleotide? Mullis immediately had the answer: the result would be the specific amplification of the DNA sequence between the two oligonucleotides. Familiar with computer programming and with the "loops" that are often used in programs, Mullis quickly realized that by repeating the elementary steps—hybridization, synthesis, and heating—he could amplify the DNA sequence between two oligonucleotides. Furthermore, Mullis realized that the amplified DNA fragment would have a precise size and be bounded at each end by the oligonucleotides that had been added as primers. Nothing prevented the oligonucleotides from being very far apart, and thus from amplifying very large DNA fragments.

Mullis thought that the idea was too simple for someone else not to have thought of it first. But when he asked his colleagues, none of them had ever heard of anything like it. They could see no reason it would not work, but none of them was particularly enthusiastic, either.

Preparing the experiment took several months; using Kornberg's original articles, Mullis had to determine the optimal concentrations for the reagents, the size of the oligonucleotides, and the composition of the medium. After this long preparation, the experiment was an immediate success. The first publication announcing the development of the technique dealt with the prenatal diagnosis of sickle cell anemia, showing the practical, applied aspect of the discovery.[25]

A number of improvements have been made to the initial protocol: most important, Kornberg's enzyme has been replaced by a DNA polymerase extracted from *Thermus aquaticus* (Taq), a bacterium that lives in hot water.[26] This enzyme is not denatured by the temperatures required to separate the DNA strands after the elongation phase. It is thus no longer necessary to add the polymerase at the beginning of each elongation phase. This led to the development of automatic machines that could be programmed to reach the required temperatures for oligonucleotide hybridization, elongation, and the denaturing of the synthesized strands. Furthermore, the whole set of operations—hybridization and elongation—could be carried out at high temperature, thus limiting the risk of nonspecific hybridization of the oligonucleotides and increasing amplification efficiency.[27]

The discovery of PCR raises a number of points relating to both the internal functioning of science in general and the specificity of molecular biology in particular. Without doubt, this was a postmature discovery.[28] All the necessary tools for its realization had existed in the 1960s. Indeed, in the 1960s Lederberg and Kornberg had discussed the possibility of obtaining large quantities of DNA using DNA polymerase.[29]

In an article, H. G. Khorana and his colleagues had gone even further in suggesting that replication should be carried out using short complementary oligonucleotides from each of the two strands of DNA.[30] At the end of the article, they outlined the three phases of PCR—hybridization with oligonucleotides, elongation by polymerase, and denaturation of the synthesized molecules—and the idea of repeating the process many times over. But there was a fundamental difference between this proposal and Mullis's idea. Khorana's aim was to copy a well-characterized DNA molecule *in vitro*. Mullis's goal was to amplify a molecule of DNA sufficiently to be able to characterize it: this fundamental difference in the aim of the experiment completely altered the importance of the procedure.

Strangely enough, Mullis's idea met with a polite but unenthusiastic response. In fact, the discovery of PCR was really a discovery only because it enabled DNA to be amplified. Many factors—nonspecific oligonucleotide fixation, unexpected stoppage of the DNA polymerase, errors in nucleotide incorporation—could have made the technique ineffective. The first presentation of results obtained with PCR was relatively modest. It was the use of Taq polymerase that made PCR sufficiently simple and effective to democratize the DNA sequence and to allow "the practice of molecular biology without a permit."[31] Mullis's colleagues, who were prudent about the possibility of developing the technique, were not being blind but simply realistic: how many apparently revolutionary discoveries have ended up in the trash can of history?

It is also important to note that PCR is perfectly representative of molecular biology and could virtually symbolize the discipline. It is a simple technique that uses biological properties—in this case, enzymes that replicate DNA—as research tools.

Two other characteristics make the discovery of PCR emblematic. The name—polymerase chain reaction—was chosen not at random but because of its reference to nuclear chain reactions.[32] In both its spirit and the person of some of its founders, molecular biology is the descendant of the physics of the 1930s and 1940s. It is also the sister of computing: the PCR protocol, with its loops of repetitive operations, is analogous to the methods of programming. PCR shares with these methods the simplicity of a series of elementary steps. Its potential, like that of a computer, is simply the product of the monotonous repetition of these steps.

The principle of PCR is so simple that, when Mullis was awarded the Nobel Prize for Chemistry in 1993, a previous winner remarked that it was a mere technical trick, without the intellectual richness that should be expected from a Nobel Prize–winning study. Nevertheless, PCR, more than any other technique, has changed the work of molecular biologists. And, after all, if it was simply a "trick," why had no one discovered it sooner?[33]

CHAPTER 21

Molecular Biology in the Life Sciences

The question of the place of molecular biology in the life sciences cannot be separated from the problem of reductionism: has molecular biology helped to reduce the functioning of organisms to the properties of the molecules that constitute them?[1] Has it therefore weakened the traditional biological disciplines?

In keeping with the spirit of the rest of the book, this question will be approached historically. Throughout the present work, several possible paths of analysis have already been highlighted. The historical approach should prevent us from straying into "metaphysical" speculations on the problem of reductionism.

As we have seen, physicists played an important role in the birth and development of molecular biology. In the 1960s, once the principal "dogmas" of molecular biology were firmly established, some molecular biologists launched a frontal attack on traditional biological disciplines, imposing a molecular vision of the phenomena that were studied in these fields. In both neurophysiology and the biology of development, these attempts failed (see Chapter 15). Today, it is noteworthy that those traditional biological disciplines that felt themselves to be threatened have since found a new lease on life. Two examples that will be dealt with here are representative of what happened to all these disciplines.

The first example, classical genetics, appeared doomed. All the phenomena it dealt with could be explained at a molecular level; the gene could be replaced by a fragment of DNA, whereas mutations could be understood as modifications of the nucleotide sequence. Several philosophers of science had shown that classical genetics was contained within molecular genetics and could be totally deduced from the new science. Nevertheless, this "encasing" of the old genetics in the new did require some minor adjustments.[2]

The paradox, however, is that despite all this, classical genetics did not disappear. For example, current research on the isolation and characterization of the genes responsible for various genetic diseases aims to use the tools of genetic engineering to clone and sequence the genes involved, deduce their structure, and imagine the function of the proteins for which they code. But the first step in this research falls within the domain of classical genetics: looking in the family of the affected individual for a genetic link between the gene responsible for the disease and other genes that have been previously localized on the chromosomes, and the estimation of the genetic distance (see Chapter 1) between these genes. The distinction between classical genetics and molecular genetics is not clear-cut: genetic markers are increasingly nucleotide sequences instead of "genes" in the classic sense of the word. Furthermore, once the study has attained a sufficient degree of resolution, the genetic map is replaced by a physical map, and intergenic distances are expressed in kilobases (thousands of base pairs). The classical and the molecular approaches thus have complementary roles.

The second example of a successful cohabitation with molecular biology is cell biology. No other discipline—apart from genetics—could have felt more threatened. After all, the molecular biologists thought that intracellular structures were simply the result of assembling macromolecules.

But without a doubt, the 1980s were the golden age of cell biology. How was this renewal possible in such a difficult context? Perhaps in part it can be explained by the development of very efficient but simple methods for studying the cell, one of which—immunofluorescence—has revealed the architecture of the cell and the presence of a cytoskeleton. Because it is simple to use, immunofluorescence does not require any great skill—unlike the electron microscope.

Nevertheless, such new techniques would not have saved cell biology if they had not unexpectedly revealed that the intracellular "traffic" was very rich. The transport of proteins between the cell surface and the various intracellular organelles is carried out by a set of vesicles. In their amino acid sequences, proteins carry signals that enable them to be taken up by these different transport systems.

Of course, the work of cell biologists, like that of geneticists, will carry on until all the proteins and enzymes involved in this intracellular traffic have been isolated, characterized, cloned, and sequenced. But intracellular events are understood not at the molecular level but at an intermediate level of analysis, in terms of compartments and vesicles.[3] As in the case of genetics, there is a complementarity between classical and molecular approaches. The molecular vision does not replace previous visions but puts them in a new light.

The biologist and philosopher of science Francisco Ayala has distinguished three different meanings of the word "reductionism."[4] First, there is ontological reductionism: the "philosophical" conviction that everything that takes place at a "higher" level of complexity follows from events that take place at a "lower" level. Contemporary biology is reductionist: biologists are convinced that all phenomena observed in organisms, no matter how complex, are the result of molecular interactions.

The second meaning of the word "reductionism" is epistemological. According to this definition, molecular biology is reductionist to the extent that it can explain, in molecular terms, observations made in other biological disciplines. The reduction of classical genetics to molecular genetics (biology) is one example of this. Harold Kincaid has shown that the same exercise was much more difficult when people tried to reduce cell biology to molecular biology.[5]

The difficulty with this epistemological view of the problem of reductionism (and probably also why it is so relatively uninteresting) is that in order to function it requires scientific knowledge to be frozen, imprisoned in rigid frameworks that do not correspond to its anarchic mode of development. What exactly is classical genetics? Can it be separated from molecular genetics without distorting history?

The final meaning of "reductionism" is more interesting because it is much more pragmatic, and much more closely linked to the real functioning of science. Ayala calls this "methodological" reductionism: the molecular approach is desirable because it is more "efficient" than other, more global approaches to biology. The members of the Rockefeller Foundation who, in the 1930s and 1940s, encouraged so many physicists to study biological problems with methods and techniques derived from modern physics, were also convinced that the reductionist approach was the most efficient for studying biological phenomena.

The supporters of methodological reductionism are often also supporters of ontological reductionism, and vice versa. It is nevertheless possible to be a supporter of ontological reductionism and thus to think that everything can be explained in terms of molecules, without being convinced that the reductionist methodology is always the best and most appropriate.

The development of molecular biology appears to be a victory for the partisans of methodological reductionism. The study of molecules and their structure has revealed some of the most detailed mechanisms of organismic functioning. But the new physical principles that Max Delbrück, and, to a lesser extent, Niels Bohr, dreamed of have not appeared. Nevertheless, when one studies closely how molecular biologists work, it becomes clear that molecular biology's methodological reductionism is only partial. As seen in the previous two examples, the concepts and models are taken from other, nonreductionist biological disciplines.

The recourse to higher levels of analysis is utterly indispensable if the edifice of modern biology is to remain intact. Only such a reference to higher levels will enable the molecular biologist (and the biochemist) to understand the finality of the biological phenomena they study, and thus to justify their research. [6]

The time has perhaps come to abandon this hierarchical vision of the sciences and to look for new kinds of relationships, on the basis of science as it is really practiced and as it really progresses.[7] From this point of view, as Harold Kincaid has pointed out, contemporary biology provides one of the best examples of the unity between different levels[8] of explanation, between different disciplines. Molecular biology did not resign itself to having to live with other biological disciplines; it entered into a far more intimate relationship with them.

As we have seen, the development of the neo-Darwinian evolutionary synthesis played a major role in the growth of molecular biology. By unifying the whole of life, neo-Darwinism supported molecular biologists in their hope of discovering fundamental principles of functioning and replication that would be shared by all organisms.

The young molecular biology quickly paid its debt to neo-Darwinism. By showing that bacteria did not adapt, but were selected, it drove Lamarckism from what had been one of its last refuges—microbiology. Nevertheless, despite the highly Darwinian declarations of the founders of molecular biology, a dispute quickly erupted between two of the founders of the modern synthesis—Theodosius Dobzhansky and Ernst Mayr—and the molecular biologists. This argument was not motivated by the population geneticists' disappointment that the limelight had shifted away from their work toward the new masters of biology.[9] Rather, they were motivated by the well-founded fear that the molecular biologists would redirect the study of evolutionary phenomena away from the individual or the population toward molecules. For the evolutionists, this was nonsensical, because even if genes were the immediate target of selection, they were expressed in individuals that interacted in populations. Evolutionary phenomena could never be explained simply at the molecular level. Furthermore, in seeking to reduce biology to physical chemistry, the molecular biologists were cancelling out the work of evolutionary biologists in trying to make biology an autonomous unified science[10] that could serve as an example to the other sciences.[11]

This was thus another example of the type of confrontation that took place in the 1970s between reductionist molecular biology and other biological disciplines. Before examining the current state of this debate, it is important to note that, irrespective of any theoretical argument, molecular biologists have provided evolutionists with some exceptional tools. While the study of evolutionary phenomena was stuck with morphology, the biochemists, followed by the molecular biologists, provided a considerable amount of new data with their studies on proteins and DNA.[12]

Molecular data confirmed that genetic variability exists within a species, as geneticists like Dobzhansky had predicted.[13] An important part of this variation was probably neutral, and was thus not subject to natural selection. This result led some geneticists, including the Japanese scientist Motoo Kimura, to propose a neutralist model of molecular evolution;[14] without being anti-Darwinian, this model nevertheless minimized the role of selection in molecular evolution.

Molecular data also enabled evolutionary biologists to travel back in time, to study the events that led to the formation of the first living cells. As seen earlier, the discovery of split genes suggested a possible mechanism for the development of complex proteins. Comparison of the nucleic acids and proteins to be found in different cell compartments supported the hypothesis that the first cells with a nucleus—eukaryotic cells—were in fact symbiotes, produced by the fusion of different acellular prokaryotic organisms, some of which gave rise to the nucleus, others to mitochondria or chloroplasts.[15]

But without doubt the most important contribution of molecular biology was the collection of vast quantities of molecular data that made it possible to make a quantitative estimate of the "genetic distance" between different species. Given that molecular variation is often neutral (for example, in most cases the substitution of the third nucleotide in a codon does not change the nature of the amino acid coded by the gene), it is possible to deduce from the genetic distance a "divergence" time for the species studied. Researchers thus have a "molecular clock" with which they can follow evolution.

The branches of the phylogenetic tree were thus redrawn on the basis of molecular data. Cladism, a new method for classifying organisms developed prior to the molecular revolution, gradually became dominant. This method requires phylogenetic distances to be estimated independent of any *a priori* idea about their relatedness.[16] The "neutral" data from molecular biology tended to support this revision of classification. This is not the place to discuss either the results that were obtained or the limits of the models that were constructed on the basis of these data. A simple reminder of the controversies surrounding the origin of mankind following the mitochondrial DNA analysis of modern human populations should be sufficient.[17] What is important to realize is that the classification of life is still being reviewed and corrected— sometimes radically—as a result of the accumulation of these new molecular data.

To return to the fundamental problem: what has molecular biology brought to the understanding of the mechanisms of evolution? The answer is short: virtually nothing, for the simple reason that the two disciplines have not interacted. Although both disciplines use the word "gene," they have not sought to bring their two meanings closer, or even to confront them. For the molecular biologist, a gene is a fragment of DNA that codes for a protein. For the population geneticist, it is a factor transmitted from generation to generation, which by its variations can confer a selective advantage (positive or negative) on the individuals carrying it.

For molecular biologists, the definition used by population geneticists appears abstract and unreal. What are these multipurpose genes that show an infinitesimal, virtually continuous variation? Sociobiologists pushed the power of genes even further and argued that there are genes responsible for altruistic behavior![18] But how can altruistic behavior be translated into molecular terms?

In the late 1970s, the evolutionary synthesis was attacked by researchers such as Stephen J. Gould and Niles Eldredge,[19] who contested the gradualist aspect of the neo-Darwinian view of evolution. They argued that the fossil record often revealed that evolution proceeded by relatively rapid bursts. Ever since Darwin, Darwinians have explained this paradox by pointing out that because of the random nature of fossilization, the fossil record is by definition incomplete. But detailed studies of fossils from sites where it is possible to make continuous observations over several million years showed that evolution did not take place in a regular manner: there were long periods of calm, during which—morphologically, at least—

fossils remained identical, followed by rapid variations.[20] This high degree of stability might suggest the existence of genetic and morphological constraints that limit the field of possible variations.[21]

For the same reasons, whereas the neo-Darwinians always explain the existence of a structure in terms of the selective advantage that it confers on the individuals that express it, Gould and Lewontin have opposed this "Panglossian" view of evolution.[22] A structure may be present for morphogenetic reasons, without *a priori* providing any selective advantage whatsoever.

The molecular biologists did not play an active part in this debate. And yet their realistic view of the role of genes was closer to the philosophy that motivated Gould and Eldredge than to that of the population geneticists. Furthermore, those molecular biologists who turned to the study of higher organisms were convinced that only an understanding of ontogeny, of embryonic development, could lead to an understanding of evolution. It was necessary to understand "the tools that the development of the embryo offers to evolutionary tinkering."[23] Surely this concept of tinkering, introduced by Jacob in 1977, suggests that the organism is the prisoner of developmental programs that it already possesses and with which it can only "tinker" for its future evolution.[24]

Given this convergence between the molecular biologists and fundamental criticisms of neo-Darwinism, it is all the more surprising that the heralds of the new discipline have remained on the sidelines of the controversy. One of the reasons for their lack of involvement was clearly the fact that their research was not sufficiently advanced to be able to make a meaningful contribution. It was only after 1980 that, thanks to the tools of genetic engineering, it became possible to isolate the genes involved in the control of embryonic development.

But this reason is not sufficient. Another factor was undoubtedly the desire of many molecular biologists, who still felt themselves to be neo-Darwinians, not to upset a theory that had contributed to the birth of their discipline and accompanied every step of its development. To attack neo-Darwinism over the problem of continuity-discontinuity was to reopen the door to heterodox theories that sought to distinguish micro- and macromutations, small evolutionary changes and massive evolutionary branching.[25] It would mean questioning the successful synthesis of the 1930s and 1940s.

It was undoubtedly this fear of rocking the foundations of their own discipline that caused many molecular biologists to shrink from the audacious models that they had begun to propose. As seen earlier, when regulatory genes were discovered in bacteria, Jacob and Monod had suggested that the complexity of higher organisms and of their embryonic development was based on the existence of a highly complex network of such regulatory genes. Indeed, the first developmental genes to be characterized—"homeobox genes"—coded for transcription factors, thus providing a striking confirmation of Jacob and Monod's views. Once regulatory genes had been discovered in 1960, it was quite reasonable to imagine that mutations affecting these genes would have important evolutionary consequences. Strangely enough, Jacob and Monod did not develop this idea, although it was later taken up, in 1975, by Mary-Claire King and Allan Wilson.[26] Following a detailed examination of proteins from chimpanzees and humans, they came to the conclusion that the structural proteins of the two species were too close to explain the obvious morphological differences between them. These differences must therefore exist at the level of regulator genes, implying that evolution is the result of mutations affecting only a part of the genome—regulator genes. Once again, this hypothesis had little impact.[27]

Up until now, molecular biology and neo-Darwinian theories of evolution have merely coexisted. Partly because of their debt to neo-Darwinism, the molecular biologists have been prudent in criticizing the more abstract aspects of evolutionary biology. Today, however, the two disciplines are beginning to interact. Molecular biologists have finally isolated and characterized a number of genes involved in development. The modification of the expression of some of these genes leads to a new morphology that is strangely similar to that of certain extinct fossil species,[28] confirming at a molecular level the link between ontogeny and phylogeny.[29] By contrast, population geneticists are turning their attention to the molecular structure and biochemical function of the genes involved in, for example, speciation.[30]

If the two disciplines come closer, will convergence take place at the expense of one of them, and in particular to the detriment of the evolutionary biologists? Will Ernst Mayr's original fear of the disappearance of those two fundamental elements of biology—the individual and the population—be realized?[31]

Several biological disciplines have been revitalized by molecular biology. Why should things be different for evolutionary biology? Such a cohabitation will require different things from the two partners: the molecular biologists will have to rediscover the complexity and richness of the living world, whereas the evolutionary biologists will have to acquire the new discipline's tools and concepts. Undoubtedly, this convergence between molecular biology and theories of evolution will be one of the major scientific events of the early twenty-first century.

Appendix: Definition of Terms

Notes

Acknowledgments

Index

Appendix: Definition of Terms

Proteins

The most important components of organisms are *proteins.* Proteins are *macromolecules,* that is, large molecules made up of several thousand atoms. Proteins have a mass of between ten thousand and several hundred thousand *daltons* (a unit of mass equal to one-twelfth of the mass of a carbon atom). Proteins are chains formed by stable chemical bonds—*covalent bonds*—between links that are called *amino acids.* Proteins are made up of twenty different amino acids, including *cysteine, proline, phenylalanine,* and *tyrosine.* Proteins are thus *polymers* formed by the *nonmonotonous* repetition of amino acids with similar but non-identical structures. Each protein is characterized by the number and nature of the amino acids that it contains, but above all by their order in the chain—the protein *sequence.*

Because of their *primary structure* (the amino acid sequence), proteins should be long, straight threads; this is not the case, however, because no sooner have they been synthesized than the chain of amino acids—the *polypeptide chain*—spontaneously folds, providing the protein with a more compact structure, which is often spherical (the *globular structure*).

Proteins have three essential functions in organisms:

1. As *enzymes,* they *catalyze* (accelerate) chemical reactions that take place within cells (the totality of these reactions forms the *metabolism*). Enzymes are extremely effective catalysts: they may accelerate a reaction more than ten billion times. There are more

than ten thousand enzymes in humans, each one specific to a particular chemical reaction. The catalytic power and the specificity of enzymes come from the fact that they form a precise complex (a *stereospecific* complex) with the molecule they transform (the *substrate*). The enzyme molecule surrounds the substrate molecule and establishes with it a series of weak chemical bonds: *van der Waals bonds, hydrogen bonds,* and *ionic bonds.*

2. Proteins also play a structural role: they contribute to the architecture of the cell and thus, indirectly, to the form of the organism. To carry out this function, they associate through a large number of weak bonds to form macromolecular complexes.
3. Proteins can also bind to genes to control their activity (see below).

Genes and DNA

An organism is thus characterized by the nature of the proteins that form it, that is, the amino acid sequence of its proteins. This information is transmitted from generation to generation by *genes.*

Human beings have between 100,000 and 200,000 genes. These genes make up the *genome* and form microscopic rodlike structures called *chromosomes.* Each human chromosome contains around 10,000 genes. In virtually all organisms, each gene is present in two copies: one from the father, the other from the mother. An organism that contains two copies of each gene is *diploid.* Organisms that contain only one copy of each gene are *haploid.* Each gene contains the information necessary for making a polypeptide chain.

Genes are made of *deoxyribonucleic acid (DNA).* DNA is formed of two molecules, each one a long chain (a *polynucleotide*) that is coiled around the other in a helix. Each polynucleotide is a polymer formed by the nonmonotonous repetition of four elementary motifs called *nucleotides* (deoxyribonucleotides).

A nucleotide contains three parts: a *phosphate group,* a *sugar* (deoxyribose), and a *base.* DNA contains four bases—*adenine, thymine, guanine,* and *cytosine.* Bases are small cyclic molecules with a relatively simple structure, each with a mass of between 100 and 200 daltons. In a polynucleotide chain the nucleotides are linked together by covalent bonds between the sugar (of one nucleotide) and the phosphate (of the next nucleotide). Each end of the polynucleotide chain is thus different, giving the chain an orientation.

In a DNA molecule, the two polynucleotide chains are in opposite orientations. Each base on one chain is paired by two or three hydrogen bonds to a base on the other chain: an adenine is always opposite a thymine, and a guanine is always opposite a cytosine. The two strands of DNA are thus *complementary:* the base sequence on one strand makes it possible to deduce the base sequence of the other.

From DNA to Proteins

The sequence of nucleotides (bases) in DNA (that is, in the gene) determines the order of amino acids in the corresponding protein: the sequence of bases *codes* for the sequence of amino acids. Each base *triplet (codon)* has a particular meaning, corresponding to a different amino acid. These correspondences are called the *genetic code.*

DNA is copied into another polynucleotide molecule called *messenger ribonucleic acid,* or *messenger RNA* (mRNA). The RNA molecule is made up of a single chain. It is complementary to the DNA strand on which it was synthesized. The passage from DNA to RNA is called *transcription.* RNA differs only slightly from DNA (the sugar is a *ribose* instead of a deoxyribose and thymine is replaced by *uracil*). The mRNA molecule is translated into a polypeptide chain.

Translation takes place on a particle formed of DNA and proteins, called a *ribosome.* (The first name given to this particle was *microsome.* Microsomes, isolated by ultracentrifugation, were a mixture of ribosomes and fragments of endoplasmic reticulum membranes. *In vivo,* the association of ribosomes with endoplasmic reticulum membranes is essential for the production of secreted proteins.) The ribosome slides along the messenger RNA. Each time the ribosome encounters a new base triplet, a small RNA called *transfer RNA* is fixed onto the messenger RNA, and the amino acid that corresponds (according to the genetic code) to this base triplet is attached to it. The different amino acids are then linked up by an enzyme present in the ribosome. The information carried by genes is thus decoded on the ribosome.

From Proteins to DNA

Some proteins bind to DNA. These proteins can play a structural role and allow the long DNA molecule to fold and form chromosomes. Some of these proteins have a more specific role: by binding on the DNA molecule just upstream of the genes, they can control the speed of transcription of DNA into RNA, that is, the *expression* of genes. These proteins are called *transcription factors.* They can act as activators or as repressors. The DNA-protein complex is called *chromatin.*

Prokaryotic and Eukaryotic Cells

All cells are surrounded by a membrane that restricts their exchanges with the external medium. There are two kinds of cells: *eukaryotic* cells, in which there are a number of intracellular structures, each isolated by a membrane and called *organelles,* and *prokaryotic* cells, which do not contain either intracellular membranes or organelles. Bacteria are prokaryotic cells. Eukaryotic cells form single-cell organisms (such as yeast) or are integrated into multicellular organisms (plants and animals).

A eukaryotic cell has a *nucleus,* which contains the chromosomes, and the *cytoplasm.* Gene transcription (DNA → RNA) takes place in the nucleus, whereas the translation of messenger RNA into protein takes place in the cytoplasm, where the ribosomes are found.

A eukaryotic cell also contains other organelles: *mitochondria,* where most of the molecules of *adenosine triphosphate (ATP),* which serve as an energy source for metabolic reactions, are produced, the *endoplasmic reticulum,* and the *Golgi apparatus*—intracellular vesicles where proteins are transported after having been synthesized, prior to being inserted into the plasma membrane or secreted. Plant cells also have *chloroplasts,* where light energy is converted into chemical energy.

Viruses are not autonomous organisms. They are formed of a genetic material (RNA or DNA) that is protected by membranes or protein structures. They hijack the cell's machinery in order to express their genetic information.

Notes

Introduction

1. This explains why, for example, this book does not deal with research on photosynthesis. Even if this topic experienced a similar "reduction" to the molecular level at around the same period as the subjects dealt with here, strictly speaking it was not part of molecular biology. For an opposite point of view, see Doris T. Zallen, "Redrawing the Boundaries of Molecular Biology: The Case of Photosynthesis," *J. Hist. Biol.,* vol. 26, 1993, pp. 65–87. By contrast, although molecular biology gives genes a central role in life and in the development of organisms, it cannot be reduced to a study of the structure and function of genes, that is, to molecular genetics. See Richard M. Burian, "Technique, Task Definition, and the Transition from Genetics to Molecular Genetics: Aspects of the Work on Protein Synthesis in the Laboratories of J. Monod and P. Zamecnik," *J. Hist. Biol.,* vol. 26, 1993, pp. 387–407.
2. The name and status of molecular biology are discussed in Robert Olby, "The Molecular Revolution in Biology," in R. C. Olby, G. N. Cantor, J. R. R. Christie, and M. J. S. Hodge (eds.), *Companion to the History of Modern Science,* Routledge, London, 1990, pp. 503–520.
3. In this book the term "paradigm" is used in a simple sense, without entering into the various debates that have ensued since its "invention" by Thomas S. Kuhn. The molecular paradigm is the new vision of life produced by molecular biology. See Thomas S. Kuhn, *The Structure of Scientific Revolutions,* University of Chicago Press, Chicago, 1970.

4. These studies will be cited wherever necessary. An almost exhaustive list of autobiographical documents can be found in Nicholas Russell, "Towards a History of Biology in the Twentieth Century: Directed Autobiographies as Historical Sources," *BJHS*, vol. 21, 1988, pp. 77–89. See also Pnina G. Abir-Am, "Noblesse Oblige: Lives of Molecular Biologists," *Isis*, vol. 82, 1991, pp. 326–343; and Jan Sapp, "Essay Review: Portraying Molecular Biology," *J. Hist. Biol.*, vol. 25, 1992, pp. 149–155.
5. Robert Olby, *The Path to the Double Helix*, Macmillan, London; 2d ed., Dover, New York, 1994.
6. Horace Freeland Judson, *The Eighth Day of Creation: The Makers of the Revolution in Biology*, Simon and Schuster, New York, 1979; 2d ed., Cold Spring Harbor Laboratory Press, 1996.
7. James D. Watson and John Tooze, *A Documentary History of Gene Cloning*, W. H. Freeman and Co., San Francisco, 1981; Stephen S. Hall, *Invisible Frontiers: The Race to Synthesize a Human Gene*, Atlantic Monthly Press, New York, 1987; Sheldon Krimsky, *Genetic Alchemy: The Social History of the Recombinant DNA Controversy*, MIT Press, Cambridge, Mass., 1982. Excluding the history of genetic engineering from that of molecular biology would be a convenient solution, but it ends up making molecular biology a "theoretical" science. It thus obscures the fact that the molecular vision of life that developed from the 1940s to the 1960s was, by its very nature, operational and applied.
8. Lily E. Kay, *The Molecular Vision of Life: Caltech, the Rockefeller Foundation and the Rise of the New Biology*, Oxford University Press, Oxford, 1993.
9. Dominique Pestre, "En guise d'introduction: quelques commentaires sur les 'temoignages oraux,'" *Cahiers pour l'histoire du CNRS*, éditions du CNRS, vol. 2, 1989, pp. 9–12.
10. In this book, the balance has at times been tilted in favor of previously neglected work, at the expense of more well known events.
11. Donald Fleming and Bernard Bailyn (eds.), *The Intellectual Migration: Europe and America, 1930–1960*, the Belknap Press of Harvard University Press, Cambridge, Mass., 1969; in particular, see Donald Fleming, "Emigré Physicists and the Biological Revolution," pp. 152–189. David Nachmansohn, *German-Jewish Pioneers in Science*, 1900–1933, Springer-Verlag, New York, 1979; Paul K. Hoch, "Migration and the Generation of New Scientific Ideas," *Minerva*, vol. 25, 1987, pp. 209–237.
12. Fernand Braudel, *La Méditerranée et le monde méditerranéen à l'époque de Philippe II*, Armand Colin, Paris, 1949.
13. Michel Tibon-Cornillot, *Les Corps transfigurés: mécanisation du vivant et imaginaire de la biologie*, Le Seuil, Paris, 1992.
14. Michel Callon and Bruno Latour, Introduction to *La Science telle qu'elle se fait: anthologie de sociologie des sciences de langue anglaise*, La Découverte, Paris, 1991, pp. 7–36.

15. Michel Morange, "Science et effet de mode," *L'État des sciences et des techniques,* Nicolas Witkowski (ed.), La Découverte, Paris, 1991, pp. 453–454.
16. Scientific publications are alleged to be belated products of scientific activity, in which "strategies" are deliberately obscured. But experience suggests the following arguments in favor of a detailed study of such publications: only such a study can unravel whole sections of the history of a science that may have been forgotten; a study of publications leads to a quantitative appreciation of the relative importance of various models or experimental approaches at a given time; publications are much less likely to be "censored" than many historians of science seem to think, especially during periods of rapid advances in knowledge—a mine of new information awaits the alert reader; finally, publications are "richer" than the work they describe: editing them is an essential part of creative scientific activity. See Frederic L. Holmes, "Scientific Writing and Scientific Discovery," *Isis,* vol. 78, 1987, pp. 220–235. In particular, writing for a public larger than fellow specialists can sometimes lead to "semantic shifts," which can be the starting point for scientific revolutions. See Christiane Sinding, "Literary Genres and the Construction of Knowledge in Biology: Semantic Shifts and Scientific Change," *Soc. Stud. Sci.,* vol. 26, 1996, pp. 43–70.
17. David Bloor, *Knowledge and Social Imagery,* Routledge and Kegan Paul, London, 1976.
18. This *a priori* position often causes historians of science to value studies that were secondary, simply because they did not lead to the development of experimental systems that could be studied reproducibly. A number of examples of such studies of "bad choices" are given below (Chapters 2 and 8).
19. The first constraint is the choice of the organism to be studied. This initial choice influences subsequent research, often in unexpected ways. See Richard M. Burian, "How the Choice of Experimental Organism Matters: Epistemological Reflections on an Aspect of Biological Practice," *J. Hist. Biol.,* vol. 26, 1993, pp. 351–367.
20. "Laboratory studies have amply shown that inanimate objects (that is, objects of scientific study) cannot be molded or made to do whatever the scientist wants. When asked to draw lines or traces on a screen, with peaks and troughs, they give very precise and definite answers. Certainly, there is room for interpretation, but the simple fact that a given apparatus is set up to produce such traces has consequences that shape all subsequent discussion. We should not ignore the fact that nonhuman scientific apparatuses also domesticate and represent reality." Callon and Latour, Introduction to *La Science telle qu'elle se fait,* p. 34.
21. The opinions expressed in this book often agree with those of the biochemist Joseph Fruton. His remarks provide historians with a stimulating set of approaches to their research. See Joseph S. Fruton, *A Skeptical Biochemist,* Harvard University Press, Cambridge, Mass., 1992.

1. The Roots of the New Science

1. Garland E. Allen, *Life Science in the Twentieth Century,* Wiley, New York, 1975; Marcel Florkin, *A History of Biochemistry,* Elsevier, Amsterdam, 1972; Joseph S. Fruton, "The Emergence of Biochemistry," *Science,* vol. 192, 1976, pp. 327–334; Claude Debru, *L'Esprit des protéines,* Hermann, Paris, 1983; P. R. Srinivasan, Joseph S. Fruton, and John T. Edsall, *The Origins of Modern Biochemistry: A Retrospect on Proteins,* the New York Academy of Science, New York, 1979; Robert E. Kohler, *From Medical Chemistry to Biochemistry,* Cambridge University Press, Cambridge, England, 1982.
2. Robert E. Kohler, "The History of Biochemistry: A Survey," *J. Hist. Biol.,* vol. 8, 1975, pp. 275–318. On the importance of this discovery, its roots and consequences, see Robert E. Kohler, "The Background to Eduard Buchner's Discovery of Cell-Free Fermentation," *J. Hist. Biol.,* vol. 4, 1971, pp. 35–61, and "The Reception of Eduard Buchner's Discovery of Cell-Free Fermentation," *J. Hist. Biol.,* vol. 5, 1972, pp. 327–353.
3. Keith J. Laidler, *The World of Physical Chemistry,* Oxford University Press, Oxford, 1993.
4. Robert Olby, *The Path to the Double Helix,* Macmillan, London, 1974, chap. 1; Debru, *L'Esprit des protéines,* chap. 2; Robert Olby, "Structural and Dynamical Explanations in the World of Neglected Dimensions," in T. J. Horder, J. A. Witkowski, and C. C. Wylie (eds.), *A History of Embryology,* Cambridge University Press, Cambridge, England, 1986, pp. 275–308; Neil Morgan, "Reassessing the Biochemistry of the 1920s: From Colloids to Macromolecules," *TIBS,* vol. 11, 1986, pp. 187–189.
5. John D. Bernal, "Structure of Proteins," *Nature,* vol. 143, 1939, pp. 663–667.
6. Debru, *L'Esprit des protéines,* chap. 2.
7. Karl Landsteiner, *The Specificity of Serological Reactions,* Charles C. Thomas, Springfield, Ill., 1936.
8. Lily E. Kay, "Molecular Biology and Pauling's Immunochemistry," *Hist. Phil. Life Sciences,* vol. 11, 1989, pp. 211–219.
9. Linus Pauling, *The Nature of the Chemical Bond,* Cornell University Press, Ithaca, 3rd ed. 1960 (1st ed. 1939); Linus Pauling, "Modern Structural Chemistry," *Science,* vol. 123, 1956, pp. 255–258; Linus Pauling, "Fifty Years of Progress in Structural Chemistry and Molecular Biology," *Daedalus,* vol. 99, 1970, pp. 988–1014; Anthony Serafini, *Linus Pauling: A Man and His Science,* Simon and Schuster, New York, 1989; Alexander Rich and Norman Davidson, *Structural Chemistry and Molecular Biology,* W. H. Freeman and Co., San Francisco, 1968; Ahmed Zewail, *The Chemical Bond: Structure and Dynamics,* Academic Press, Cambridge, Mass., 1992; Thomas Hager, *Force of Nature: The Life of Linus Pauling,* Simon and Schuster, New York, 1995.

10. Linus Pauling, "Nature of Forces between Molecules of Biological Interest," *Nature,* vol. 161, 1948, pp. 707–709. Among the weak bonds, Pauling emphasized the role of hydrogen bonds, attributing them with a higher energy value than is given today. He paid little attention to hydrophobic interactions or to what are now called changes in entropy. These errors were probably useful in the development of the concept of weak bond, because they made things simpler. See Howard Schachman, "Summary Remarks: A Retrospect on Proteins," in P. R. Srinivasan, Joseph S. Fruton, and John T. Edsall, *The Origins of Modern Biochemistry,* New York Academy of Sciences, New York, 1979, pp. 363–373.
11. Alfred E. Mirsky and Linus Pauling, "On the Structure of Native, Denatured and Coagulated Proteins," *Proc. Natl. Acad. Sci. USA,* vol. 22, 1936, pp. 439–447.
12. Alfred H. Sturtevant, *A History of Genetics,* Harper and Row, New York, 1965; Elof A. Carlson, *The Gene: A Critical History,* Saunders, Philadelphia, 1966; Garland E. Allen, *Life Science in the Twentieth Century,* Wiley, New York, 1975; Ernst Mayr, *The Growth of Biological Thought: Diversity, Evolution, and Inheritance,* Harvard University Press, Cambridge, Mass., 1982; Peter J. Bowler, *The Mendelian Revolution: The Emergence of Hereditarian Concepts in Modern Science and Society,* Johns Hopkins University Press, Baltimore, 1989; Lindley Darden, *Theory Change in Science: Strategies from Mendelian Genetics,* Oxford University Press, New York, 1991; Jean-Louis Fischer and William H. Schneider, *Histoire de la génétique, pratique, techniques et théories,* ARPEM et Sciences en situation, Paris, 1990; Robert Olby, *The Origins of Mendelism,* Chicago University Press, Chicago, 1966; 2d ed., 1985; Robert E. Kohler, *Lords of the Fly: Drosophila Genetics and the Experimental Life,* University of Chicago Press, Chicago, 1994. Even the idea that Mendel's results were forgotten and then rediscovered is disputed by contemporary historians. First, Mendel's results were not unknown, and second, the context in which they were rediscovered was very different from that in which they were first stated. See Robert Olby, "Mendel no Mendelian?" *Hist. Sci.,* vol. 17, 1979, pp. 53–72; and Augustine Brannigan, "The Reification of Mendel," *Soc. Stud. Sci.,* vol. 9, 1979, pp. 423–454.
13. Thomas H. Morgan, Alfred H. Sturtevant, Hermann J. Muller, and Calvin B. Bridges, *The Mechanism of Mendelian Heredity,* Henry Holt and Co., New York, 1915.
14. Barbara A. Kimmelman, "Agronomie et théorie de Mendel," in Fischer and Schneider, *Histoire de la génétique;* Robert Olby, "Rôle de l'agriculture et de l'horticulture britanniques," in Fischer and Schneider, *Histoire de la génétique.*
15. Bowler, *The Mendelian Revolution;* Daniel J. Kevles, "Genetics in the United States and Great Britain, 1890–1930: A Review with Speculations," *Isis,* vol. 71, 1980, pp. 441–455; Jonathan Harwood, "National Styles in Science: Genetics in Germany and the United States between the World Wars," *Isis,* vol. 78, 1987, pp. 390–414; Jonathan Harwood, *Styles of Scientific Thought: The German*

Genetics Community, 1900–1933, University of Chicago Press, Chicago, 1993; Richard M. Burian, Jean Gayon, and Doris Zallen, "The Singular Fate of Genetics in the History of French Biology, *J. Hist. Biol.,* vol. 21, 1988, pp. 357–402.

16. Harwood, "National Styles in Science"; and *Styles of Scientific Thought.*
17. Charles Lenay, *La Découverte des lois de l'hérédité. Une anthologie,* Presses Pocket, Paris, 1990; Raphael Falk, "The Struggle of Genetics for Independence," *J. Hist. Biol.,* vol. 28, 1995, pp. 219–246.
18. Hermann J. Muller, "Artificial Transmutation of the Gene," *Science,* vol. 66, 1927, pp. 84–87; Elof Axel Carlson, "An Acknowledged Founding of Molecular Biology: Hermann J. Muller's Contributions to Gene Theory, 1910–1936," *J. Hist. Biol.,* vol. 4, 1971, pp. 149–170; Elof Axel Carlson, *Genes, Radiation, and Society: The Life and Work of H. J. Muller,* Cornell University Press, Ithaca, 1982.
19. Olby, *The Path to the Double Helix,* chap. 7. DNA had been discovered in 1869 by Johann Friedrich Miescher. See Franklin H. Portugal and Jack S. Cohen, *A Century of DNA: A History of the Structure and Function of the Genetic Substance,* MIT Press, Cambridge, Mass., 1977.
20. Hermann J. Muller, "The Gene," Pilgrim Trust Lecture, *Proc. Roy. Soc. B,* vol. 134, 1947, pp. 1–37 (quotation from p. 1).
21. This is shown by the rare detailed models of genes proposed in the 1930s and 1940s. See Dorothy M. Wrinch, "Chromosome Behaviour in Terms of Protein Pattern," *Nature,* vol. 134, 1934, pp. 978–979; N. K. Koltzoff, "Les molécules héréditaires," *Actualités Scientifiques et Industrielles,* no. 776, Hermann, Paris, 1939.
22. Hermann J. Muller, "Resume and Perspectives of the Symposium on Genes and Chromosomes," *Cold Spring Harbor Symp. Quant. Biol.,* vol. 9, 1941, pp. 290–308; Olby, *The Path to the Double Helix,* chap. 7.
23. Jordan's model was part of his larger project of "quantum biology": Richard H. Beyler, "Targeting the Organism: The Scientific and Cultural Context of Pascual Jordan's Quantum Biology, 1932–1947," *Isis,* vol. 87, 1996, pp. 248–273.
24. Linus Pauling and Max Delbrück, "The Nature of the Intermolecular Forces Operative in Biological Processes," *Science,* vol. 92, 1940, pp. 77–79.
25. Allen, *Life Science;* Ernst Mayr and William B. Provine (eds.), *The Evolutionary Synthesis: Perspectives in the Unification of Biology,* Harvard University Press, Cambridge, Mass., 1980; Mayr, *The Growth of Biological Thought;* Fischer and Schneider, *Histoire de la génétique;* Jean Gayon, *Darwin et l'après-Darwin. Une histoire de l'hypothèse de sélection naturelle,* Kimé, Paris, 1992 (translated as *Darwinism's Struggle for Survival: Heredity and the Hypothesis of Natural Selection,* Cambridge University Press, Cambridge, England, forthcoming); V. B. Smocovitis, "Unifying Biology: The Evolutionary Synthesis and Evolutionary Biology," *J. Hist. Biol.,* vol. 25, 1992, pp. 1–65.

26. Theodosius Dobzhansky, *Genetics and the Origin of Species,* Columbia University Press, New York, 1937; George G. Simpson, *Tempo and Mode in Evolution,* Columbia University Press, New York, 1944. Dobzhansky played a key role in confronting the models of Sewall Wright, J. B. S. Haldane, and R. A. Fisher with experimental data, thus helping to reconcile field scientists and theoretical biologists. See Robert E. Kohler, "Drosophila and Evolutionary Genetics: The Moral Economy of Scientific Practice," *Hist. Sci.,* vol. 24, 1991, pp. 335–375.

2. The One Gene–One Enzyme Hypothesis

1. George W. Beadle and Edward L. Tatum, "Genetic Control of Biochemical Reactions in Neurospora," *Proc. Natl. Acad. Sci. USA,* vol. 27, 1941, pp. 499–506.
2. Harriet Zuckerman and Joshua Lederberg, "Postmature Scientific Discovery?" *Nature,* vol. 324, 1986, pp. 629–631.
3. Robert Olby, *The Path to the Double Helix,* Macmillan, London; Krishna R. Dronamraju, "Profiles in Genetics: George Wells Beadle and the Origin of the Gene-Enzyme Concept," *J. Hered.,* vol. 82, 1991, pp. 443–446; Arnold W. Ravin, "The Gene as Catalyst; the Gene as Organism," *Stud. Hist. Biol.,* vol. 1, 1977, pp. 1–45.
4. Alexander G. Bearn, *Archibald Garrod and the Individuality of Man,* Clarendon Press, Oxford, 1993.
5. Cited in Olby, *The Path to the Double Helix,* p. 130.
6. Archibald Garrod, *The Inborn Errors of Metabolism,* Oxford University Press, London, 1909; new ed. by H. Harris, London, 1963.
7. Ephrussi and Beadle's work, which used the embryological method of transplantation, was one of the first attempts to reconcile genetics and embryology. Beadle and Ephrussi continued to work on this system for many years. They tested the effect of nutrition on eye color in order to identify the substance responsible for pigment formation. Beadle's subsequent change of approach can be explained by the direction taken by his initial work with Ephrussi. The German chemist Adolph Butenandt was the first to determine the nature of the pigment. See Robert E. Kohler, "Systems of Production: Drosophila, Neurospora and Biochemical Genetics," *HSPS,* vol. 22, 1991, pp. 87–130. For a study of Ephrussi's career, see Richard M. Burian, Jean Gayon, and Doris T. Zallen, "Boris Ephrussi and the Synthesis of Genetics and Embryology," in *Developmental Biology,* vol. 7, *A Conceptual History of Modern Embryology,* Scott F. Gilbert (ed.), Plenum Press, New York, 1991, pp. 207–227. Jean Gayon has shown that Ephrussi explained his data by a

model very different from that later proposed by Beadle and Tatum. See Jean Gayon, "Génétique de la pigmentation de l'oeil de la drosophile: la contribution spécifique de Boris Ephrussi," in *Les sciences biologiques et médicales en France: 1920–1950,* Claude Debru, Jean Gayon, and Jean-François Picard (eds.), CNRS Editions, Paris, 1994, pp. 187–206.

8. Garland E. Allen, *Thomas Hunt Morgan: The Man and His Science,* Princeton University Press, Princeton, 1978; E. A. Carlson, *The Gene: A Critical History,* Saunders, Philadelphia, 1966; Peter J. Bowler, *The Mendelian Revolution: The Emergence of Hereditarian Concepts in Modern Science and Society,* Johns Hopkins University Press, Baltimore, 1989.
9. Richard B. Goldschmidt, *Physiological Genetics,* McGraw-Hill, New York, 1938; Garland E. Allen, "Opposition to the Mendelian-Chromosome Theory: The Physiological and Developmental Genetics of Richard Goldschmidt," *J. Hist. Biol.,* vol. 7, 1974, pp. 49–92.
10. Scott F. Gilbert, "Induction and the Origins of Developmental Genetics," in *Developmental Biology;* Rose-Marie Ponsot, "C. H. Waddington: l'évolution d'un évolutionniste," diss., Université Paris-I, Paris, 1987; Scott F. Gilbert, "Cellular Politics: Ernest Everett Just, Richard B. Goldschmidt, and the Attempt to Reconcile Embryology and Genetics," in *The American Development of Biology,* Ronald Rainger, Keith R. Benson, and Jane Maienschein (eds.), University of Pennsylvania Press, Philadelphia, 1988, chap. 10, pp. 311–346.
11. Scott F. Gilbert, "The Embryological Origins of the Gene Theory," *J. Hist. Biol.,* vol. 11, 1978, pp. 307–351.
12. Jane Maienschein, "What Determines Sex? A Study of Converging Approaches, 1880–1916," *Isis,* vol. 75, 1984, pp. 457–480; Muriel Lederman, "Research Note: Genes on Chromosomes—The Conversion of Thomas Hunt Morgan, *J. Hist. Biol.,* vol. 22, 1989, pp. 163–176.
13. Jan Sapp, "The Struggle for Authority in the Field of Heredity, 1900–1932: New Perspectives on the Rise of Genetics," *J. Hist. Biol.,* vol. 16, 1983, pp. 311–342.
14. Allen, *Thomas Hunt Morgan,* chap. 9.
15. Lily E. Kay, "Selling Pure Science in Wartime: The Biochemical Genetics of G. W. Beadle," *J. Hist. Biol.,* vol. 22, 1989, pp. 73–101; Robert E. Kohler, "Systems of Production: Drosophila, Neurospora and Biochemical Genetics," *HSPS,* vol. 22, 1991, pp. 87–130; George W. Beadle, "Recollections," *Ann. Rev. Biochem.,* vol. 43, 1974, pp. 1–3.
16. Lily E. Kay, "Selling Pure Science in Wartime"; Lily E. Kay, "Microorganisms and Macromanagement," in *The Molecular Vision of Life: Caltech, the Rockefeller Foundation, and the Rise of the New Biology,* Oxford University Press, Oxford, 1993. The microbiological determination of amino acids was already a "classic" method. See, for instance, the review by Esmond E. Snell, "The Microbiological Assay of Amino Acids," *Adv. Prot. Chem.,* vol. 2, 1945, pp. 85–118.

17. George W. Beadle, "Genetic Control of Biochemical Reactions," *Harvey Lect.*, vol. 40, 1945, p. 193.
18. Ernest P. Fischer and Carol Lipson, *Thinking about Science: Max Delbrück and the Origin of Molecular Biology*, W. W. Norton, New York, 1988, pp. 169–173.
19. Olby, *The Path to the Double Helix*, chap. 9; Ravin, "The Gene as Catalyst," pp. 1–45.
20. Jan Sapp, *Where the Truth Lies: Franz Moewus and the Origins of Molecular Biology*, Cambridge University Press, Cambridge, England, 1990; Jan Sapp, "What Counts as Evidence, or Who Was Franz Moewus and Why Was Everybody Saying Such Terrible Things about Him," *Hist. Phil. Life Sci.*, vol. 9, 1987, pp. 277–308.
21. Sapp, "What Counts as Evidence," p. 307.
22. Lily E. Kay noted that Moewus was strangely missing from Sapp's book (Lily E. Kay, "Book Reviews," *Isis*, vol. 83, 1992, pp. 160–161). One could say almost the same thing about Moewus's research.

3. The Chemical Nature of the Gene

1. Oswald T. Avery, Colin MacLeod, and Maclyn McCarty, "Studies on the Chemical Nature of the Substance Inducing Transformation of Pneumococcal Types," *J. Exp. Med.*, vol. 79, 1944, pp. 137–158.
2. Alfred D. Hershey and Martha Chase, "Independent Functions of Viral Proteins and of Nucleic Acids in the Growth of the Bacteriophage," *J. Gen. Physiol.*, vol. 36, 1952, pp. 39–56.
3. Gunther S. Stent, "Prematurity and Uniqueness in Scientific Discovery," *Scientific American*, vol. 228, 1972, pp. 84–93.
4. H. V. Wyatt, "When Does Information Become Knowledge," *Nature*, vol. 235, 1972, pp. 86–89.
5. René J. Dubos, *The Professor, the Institute, and DNA*, Rockefeller University Press, New York, 1976; Olga Amsterdamska, "Between Medicine and Science: The Research Career of Oswald T. Avery," in Ilana Löwy (ed.), *Medicine and Change: Historical and Sociological Studies of Medical Innovation*, INSERM, J. Libbey Eurotext, 1992, pp. 181–212; Olga Amsterdamska, "From Pneumonia to DNA: The Research Career of Oswald T. Avery," *Hist. Stud. Phys. Biol. Sci.*, vol. 24, 1993, pp. 1–40.
6. Maclyn McCarty, *The Transforming Principle*, W. W. Norton Company, New York, 1985.
7. Avery, MacLeod, and McCarty, "Studies on the Chemical Nature of the Substance" *J. Exp. Med.*, vol. 79, 1944, pp. 137–158.
8. Dubos, *The Professor*.
9. Wyatt, "When Does Information Become Knowledge," pp. 86–89.

10. Horace F. Judson, *The Eighth Day of Creation: The Makers of the Revolution in Biology,* Simon and Schuster, New York, 1979, p. 60.
11. Robert Olby, *The Path to the Double Helix,* Macmillan, London, chap. 13.
12. François Jacob, *La Logique du Vivant,* Gallimard, Paris, p. 282 (translated as *The Logic of Life,* Pantheon Press, New York, 1974).
13. Olby, *The Path to the Double Helix,* chap 9.
14. Max Delbrück, "A Theory of Autocatalytic Synthesis of Polypeptide and Its Application to the Problem of Chromosome Reproduction," *Cold Spring Harbor Symp. Quant. Biol.,* vol. 9, 1941, pp. 122–124.
15. Dubos, *The Professor.* Some historians have noted that even in his 1944 paper, Avery described the transforming principle, not as a gene, but as a specific mutagenic agent (Bernardino Fantini, "Genes and DNA," *Hist. Phil. Life Sci.,* vol. 10, 1988, pp. 145–151). The distinction between gene and specific mutagenic agent, so important for modern genetics, was not at all clear for Avery and his contemporaries, for whom genes and mutagens had the same "catalytic" action.
16. H. V. Wyatt, "Knowledge and Prematurity: The Journey from Transformation to DNA," *Perspec. Biol. Med.,* vol. 18, 1975, pp. 149–156.
17. Quoted by Ilana Löwy, "Variances in Meaning in Discovery Accounts: The Case of Contemporary Biology," *HSPS,* vol. 21, 1990, pp. 87–121 (quote from p. 112).
18. Michel Morange, "La révolution silencieuse de la biologie moléculaire," *Débat,* vol. 10, 1982, pp. 62–75.
19. Colin MacLeod felt that the subject was so problematic that it would be difficult to publish any study of transformation. For three years he turned to a topic that was more "rewarding" in terms of publication (see McCarty, *The Transforming Principle,* pp. 96–100).
20. Nicholas Russell, "Oswald Avery and the Origin of Molecular Biology," *BJHS,* vol. 21, 1988, pp. 393–400.
21. Erwin Chargaff, *Heraclitean Fire: Sketches from a Life before Nature,* Rockefeller University Press, New York, 1978, pp. 86–100. Chargaff's work is also described in Olby, *The Path to the Double Helix,* pp. 211–219.
22. Rollin D. Hotchkiss, "The Identification of Nucleic Acids as Genetic Determinants," in P. R. Srinivasan, Joseph S. Fruton, and John T. Edsall, *The Origins of Modern Biochemistry,* New York Academy of Sciences, New York, 1979, pp. 321–342.

4. The "Phage Group"

1. The publication of a festschrift in honor of Max Delbrück provided a belated baptism for the group. It was closely followed in 1969 by the attribution of the Nobel Prize to its three founders. See John Cairns, Gunther S. Stent, and

James D. Watson (eds.), *Phage and the Origins of Molecular Biology,* Cold Spring Harbor Laboratory Press, Cold Spring Harbor, 1966 (expanded ed., 1992); Nicholas C. Mullins, "The Development of a Scientific Specialty: The Phage Group and the Origins of Molecular Biology," *Minerva,* vol. 10, 1972, pp. 51–82; D. Fleming, "Emigré Physicists and the Biological Revolution," *Perspec. Am. Hist.,* vol. 2, pp. 176–213, reprinted in Donald Fleming and Bernard Bailyn (eds.), *The Intellectual Migration: Europe and America, 1930–1960,* the Belknap Press of Harvard University Press, Cambridge, Mass., 1969.

2. Lily E. Kay, "Conceptual Models and Analytical Tools: The Biology of Physicist Max Delbrück," *J. Hist. Biol.,* vol. 18, 1985, pp. 207–246; Ernst P. Fischer and Carol Lipson, *Thinking about Science: Max Delbrück and the Origins of Molecular Biology,* W. W. Norton, New York, 1988; Thomas D. Brock, *The Emergence of Bacterial Genetics,* Cold Spring Harbor Laboratory Press, Cold Spring Harbor, 1990, chap. 6.
3. Niels Bohr, "Light and Life," *Nature,* vol. 131, 1933, pp. 421–423 and 457–459.
4. Nikolaï W. Timofeeff-Ressovsky, Karl G. Zimmer, and Max Delbrück, "Über die Natur der Genmutation und der Genstruktur," *Nachr. Ges. Wiss. Göttingen math-phys. Kl.,* vol. 6, 1935, pp. 190–245. The life of Nikolaï Timofeeff-Ressovsky illustrates the difficulties faced by all those who could not or would not choose the right camp in the Second World War. Having left Russia in 1926 to work with the German psychiatrist and neurophysiologist Oskar Vogt at the end of his visit to the USSR, Timofeeff-Ressovsky stayed on to work in Germany during the war. At the end of the war, he was sent to a labor camp in the Soviet Union. Frédéric Joliot-Curie's intervention and Timofeef-Ressovsky's scientific abilities helped to set him free, after which he was put in charge of a radiation biology laboratory in Sverdlovsk, in the Urals. He was later charged with having participated in research on prisoners during the war and of having tested the effect of radioactive compounds on them. This alleged collaboration with the Nazis—if it happened—did not enable him to rescue his son, an anti-Nazi resistant, from the Mauthausen concentration camp, where he died. Timofeeff-Ressovsky's scientific contribution was very important. In addition to his collaboration with Delbrück, he made several fundamental discoveries in population genetics. He helped make Chetverikov's Russian school known, and hence played a major role in reconciling genetics and evolutionary biology. His influence seems to have been particularly strong on Dobzhansky. See Diane B. Paul and Costas B. Krimbas, "Nikolaï W. Timofeeff-Ressovsky," *Scientific American,* vol. 266, Feb. 1992, pp. 64–70; Max F. Perutz, "Erwin Schrödinger's 'What Is Life?' and Molecular Biology," in *Schrödinger: Centenary Celebration of a Polymath,* C. W. Kilmister (ed.), Cambridge University Press, Cambridge, England, 1987; Bentley Glass, "Timofeeff-Ressovsky, Nikolaï Wladimirowich," F. L. Holmes (ed.), *Dictionary of Scientific Biographies,* vol. 18, suppl. II, Charles Scribner's Sons, 1990.

5. Keith L. Manchester, *Exploring the Gene with X-Rays, TIG,* vol. 12, 1996, pp. 515–518.
6. Erwin Schrödinger, *What Is Life?* Cambridge University Press, Cambridge, England, 1944.
7. Salvador E. Luria, *A Slot Machine, a Broken Test Tube: An Autobiography,* Harper and Row, New York, 1984.
8. Max Delbrück and Nikolaï W. Timofeeff-Ressovsky, "Cosmic Rays and the Origin of Species," *Nature,* vol. 137, 1936, pp. 358–359.
9. Elof A. Carlson, *The Gene: A Critical History,* Saunders, Philadelphia, 1966.
10. Lily E. Kay, "Quanta of Life: Atomic Physics and the Reincarnation of Phage," *Hist. Phil. Life Sci.,* vol. 14, 1992, pp. 3–21.
11. Étienne Klein, *Conversations avec le Sphinx. Les paradoxes en physique,* Albin Michel, Paris, 1991.
12. William C. Summers, "How Bacteriophage Came to Be Used by the Phage Group," *J. Hist. Biol.,* vol. 26, 1993, pp. 255–267.
13. The bacteriophage is one of the best examples of the importance of the "right tool for the job." Adèle E. Clark and Joan H. Fujimura, *The Right Tools for the Job: At Work in Twentieth-Century Life Sciences,* Princeton University Press, Princeton, N.J., 1992.
14. Lily E. Kay, "Virus, Enzyme ou gène? Le problème du bactériophage (1917–1947)," in *L'Institut Pasteur. Contributions à son histoire,* La Découverte, Paris, 1991.
15. Ton Van Helvoort, "The Construction of Bacteriophage as Bacterial Virus: Linking Endogenous and Exogenous Thought Styles," *J. Hist. Biol.,* vol. 27, 1994, pp. 91–139.
16. Kay, "Virus, Enzyme ou gène?" Ton Van Helvoort, "The Controversy between John H. Northrop and Max Delbrück on the Formation of Bacteriophage: Bacterial Synthesis or Autonomous Multiplication," *Annals of Science,* vol. 49, pp. 545–575.
17. Emory Ellis and Max Delbrück, "The Growth of Bacteriophage," *J. Gen. Physiol.,* vol. 22, 1939, pp. 365–384.
18. Mullins, "The Development of a Scientific Specialty."
19. Ernest P. Fischer and Carol Lipson, *Thinking about Science: Max Delbrück and the Origin of Molecular Biology,* W. W. Norton, New York, 1988, p. 147.
20. Salvador E. Luria and Thomas F. Anderson, "The Identification and Characterization of Bacteriophages with the Electron Microscope," *Proc. Natl. Acad. Sci. USA,* vol. 28, 1942, pp. 127–130.
21. H. Ruska, "Über ein neues bei der bakteriophagen Lyse auftretendes Formelement," *Naturwiss.,* vol. 29, 1941, pp. 367–368.
22. Luria, *A Slot Machine.*
23. Seymour S. Cohen, "Synthesis of Bacterial Viruses, Synthesis of Nucleic Acid

and Protein in *Escherichia coli* Infected with T2 r^+ Bacteriophage," *J. Biol. Chem.*, vol. 174, 1948, pp. 295–303.

24. After the war, the phage system became very well known, even outside biological circles. When John von Neumann wrote to Norbert Wiener in 1946 to discuss the direction taken by research in cybernetics and the functioning of the human brain, he quoted the example of the bacteriophage as a model for studies aiming at understanding elementary biological functioning. (Pesi R. Masani, *Norbert Wiener, 1894–1964,* Vita Mathematica, vol. 5, Birkhaüser Verlag, Basel, 1990, pp. 242–249.)
25. Gunther S. Stent, "Max Delbrück," *TIBS,* vol. 6, 1981, pp. iii–iv.
26. Fischer and Lipson, *Thinking about Science,* p. 179.
27. An important spin-off of the work of the phage group, in a different branch of science, was the development by Renato Dulbecco, on Delbrück's recommendation, of a new method of quantitatively studying animal viruses. This method, now widespread, was derived from the method used for studying bacteriophages (see Chapter 15).
28. Alfred D. Hershey and Martha Chase, "Independent Functions of Viral Protein and Nucleic Acid in Growth of Bacteriophage," *J. Gen. Physiol.,* vol. 36, 1952, pp. 39–56.
29. Gunther S. Stent, "That Was the Molecular Biology That Was," *Science,* vol. 160, 1968, pp. 390–395.
30. H. V. Wyatt, "How History Has Blended," *Nature,* vol. 249, 1974, pp. 803–805.
31. H. V. Wyatt, "Knowledge and Prematurity: The Journey from Transformation to DNA," *Perspec. Biol. Med.,* vol. 18, 1975, pp. 149–156.
32. André Boivin, Roger Vendrely, and Colette Vendrely, "L'acide désoxyribonucléique du noyau cellulaire, dépositaire des caractères héréditaires; arguments d'ordre analytique," *CRAS,* Paris, vol. 226, 1948, pp. 1061–1063.
33. Roger Vendrely and Colette Vendrely, "La teneur du noyau cellulaire en acide désoxyribonucléique à travers les organes, les individus et les espèces animales," *Experientia,* vol. 4, 1948, pp. 434–436.

5. The Birth of Bacterial Genetics

1. William Bulloch, *The History of Bacteriology,* Oxford University Press, 1938; (reprinted) Dover Publications Inc., New York, 1979.
2. Olga Amsterdamska, "Stabilizing Instability: The Controversy over Cyclogenic Theories of Bacterial Variation during the Interwar Period," *J. Hist. Biol.,* vol. 24, 1991, pp. 191–222.

3. Robert E. Kohler, "Innovation in Normal Science: Bacterial Physiology," *Isis,* vol. 76, 1985, pp. 162–181.
4. V. B. Smocovitis, "Unifying Biology: The Evolutionary Synthesis and Evolutionary Biology," *J. Hist. Biol.,* vol. 25, 1992, pp. 1–65; Theodosius G. Dobzhansky, *Genetics and the Origin of Species,* Columbia University Press, New York, 1937.
5. Cyril Hinshelwood, *The Chemical Kinetics of the Bacterial Cell,* the Clarendon Press, Oxford, 1946.
6. William Summers, "From Culture as Organism to Organism as Cell: Historical Origins of Bacterial Genetics," *J. Hist. Biol.,* vol. 24, 1991, pp.171–190.
7. I. M. Lewis, "Bacterial Variation with Special Reference to Behavior of Some Mutabile Strains of Colon Bacteria in Synthetic Media," *J. Bacteriol.,* vol. 26, 1934, pp. 619–639.
8. Salvador E. Luria and Max Delbrück, "Mutations of Bacteria from Virus Sensitivity to Virus Resistance," *Genetics,* vol. 28, 1943, pp. 491–511.
9. Salvador E. Luria and Max Delbrück, "Interference between Bacterial Viruses: 1-Interference between Two Bacterial Viruses Acting Upon the Same Host, and the Mechanism of Virus Growth," *Arch. Bioch.,* vol. 1, 1942, pp. 111–141.
10. Salvador E. Luria, *A Slot Machine, a Broken Test Tube: An Autobiography,* Harper and Row, New York, 1984.
11. Ibid., p. 20.
12. Eugene Wollman, Fernand Holweck, and Salvador Luria, "Effect of Radiations on Bacteriophage C16," *Nature,* vol. 145, 1940, pp. 935–936.
13. S. E. Luria and T. F. Anderson, "The Identification and Characterization of Bacteriophages with the Electron Microscope," *Proc. Natl. Acad. Sci. USA,* vol. 28, 1942, pp. 127–130.
14. Luria, *A Slot Machine,* p. 75.
15. Ernest P. Fischer and Carol Lipson, *Thinking about Science: Max Delbrück and the Origins of Molecular Biology,* W. W. Norton, New York, 1988, pp. 142–147.
16. Salvador E. Luria, "Recent Advances in Bacterial Genetics," *Bacteriol. Rev.,* vol. 11, 1947, pp. 1–40. Recent experiments have limited the generality of Luria and Delbrück's results, without calling them into question. If bacteria are subject to a selection pressure less drastic than infection by a bacteriophage, they can increase their mutation rate, thus adapting to new growth conditions (John Cairns, Julie Overbaugh, and Stephan Miller, "The Origin of Mutants," *Nature,* vol. 335, 1988, pp. 142–145). For a discussion of these recent results, see Evelyn Fox Keller, "Between Language and Science: The Question of Directed Mutation in Molecular Genetics," *Perspec. Biol. Med.,* vol. 35 (2), 1992, pp. 293–306.
17. René J. Dubos, *The Bacterial Cell in Its Relation to Problems of Virulence, Immunity, and Chemotherapy,* Harvard University Press, Cambridge, Mass., 1945, p. 176 et seq.

18. Harriet Zuckerman and Joshua Lederberg, "Postmature Scientific Discovery?" *Nature,* vol. 324, 1986, pp. 629–631. Dobzhansky, one of the key participants in the evolutionary synthesis, played an important role in this unification of biology. He helped circulate information among geneticists, biochemists, and physicists, and in particular, he made Avery's findings known to geneticists.
19. Peter Medawar, *The Art of the Soluble,* Methuen, London, 1967.
20. Joshua Lederberg, "A Fortieth Anniversary Reminiscence," *Nature,* vol. 324, 1986, pp. 627–628; "Genetic Recombination in Bacteria: A Discovery Account," *Ann. Rev. Genet.,* vol. 21, 1987, pp. 23–46.
21. Joshua Lederberg and Edward L. Tatum, "Novel Genotypes in Mixed Cultures of Biochemical Mutants of Bacteria," *Cold Spring Harbor Symp. Quant. Biol.,* vol. 11, 1946, pp. 113–114. These results were published in a short paper sent to *Nature:* Joshua Lederberg and Edward L. Tatum, "Gene Recombination in *Escherichia coli,*" *Nature,* vol. 158, 1946, p. 558.
22. Thomas D. Brock, *The Emergence of Bacterial Genetics,* Cold Spring Harbor Laboratory Press, Cold Spring Harbor, 1990.
23. Salvador E. Luria, "Mutations of Bacterial Viruses Affecting Their Host Ranges," *Genetics,* vol. 30, 1945, pp. 84–99.
24. Alfred D. Hershey, "Mutation of Bacteriophage with Respect to Type of Plaque," *Genetics,* vol. 31, 1946, pp. 620–640.
25. Alfred D. Hershey, "Spontaneous Mutations in Bacterial Viruses," *Cold Spring Harbor Symp. Quant. Biol.,* vol. 11, 1946, pp. 67–77; M. Delbrück and W. T. Bailey, "Induced Mutations in Bacterial Viruses," *Cold Spring Harbor Symp. Quant. Biol.,* vol. 11, 1946, pp. 33–37.
26. J. Lederberg, E. M. Lederberg, N. D. Zinder, and E. R. Lively, "Recombination Analysis of Bacterial Heredity," *Cold Spring Harbor Symp. Quant. Biol.,* vol. 16, 1951, pp. 413–443.
27. William Hayes, "Recombination in Bact. coli K12: Unidirectional Transfer of Genetic Material," *Nature,* vol. 169, 1952, pp. 118–119.
28. Elie L. Wollman, "Bacterial Conjugation," pp. 216–225, in John Cairns, Gunther S. Stent, and James D. Watson (eds.), *Phage and the Origins of Molecular Biology,* Cold Spring Harbor Laboratory Press, Cold Spring Harbor, 1966 (expanded ed., 1992).
29. Elie L. Wollman and François Jacob, "Sur le mécanisme de transfert du matériel génétique au cours de la recombinaison chez *Escherichia coli* K12," *Comptes rendus de l'Académie des sciences,* vol. 240, 1955, pp. 2449–2451.
30. Lederberg, Lederberg, Zinder, and Lively, "Recombination Analysis of Bacterial Heredity."
31. Lederberg, "Genetic Recombination in Bacteria," pp. 33–34.
32. Norton D. Zinder and Joshua Lederberg, "Genetic Exchange in *Salmonella,*" *J. Bacteriol.,* vol. 64, 1952, pp. 679–699.

33. Milislav Demerec and P. E. Hartman, "Tryptophan Mutants in *Salmonella typhimurium*," pp. 5–33, and P. E. Hartman, "Linked Loci in the Control of Consecutive Steps in the Primary Pathway of Histidine Synthesis in *Salmonella typhimurium*," pp. 35–61, in *Genetic Studies in Bacteria*, Carnegie Institution of Washington Publication, 612, Washington, D.C., 1956.
34. M. L. Morse, E. M. Lederberg, and J. Lederberg, "Transduction in *Escherichia coli* K12," *Genetics*, vol. 41, 1956, pp. 142–156.

6. The Crystallization of the Tobacco Mosaic Virus

1. Hermann J. Muller, "Variations Due to Change in the Individual Gene," *Amer. Natur.*, vol. 22, 1922, pp. 32–50.
2. Ton Van Helvoort, "What Is a Virus? The Case of Tobacco Mosaic Disease," *Stud. Hist. Phil. Sci.*, vol. 22, 1991, pp. 557–588.
3. Robert Olby, *The Path to the Double Helix*, Macmillan, London, 1974, chap. 10.
4. Lily E. Kay, "W. M. Stanley's Crystallization of the Tobacco Mosaic Virus, 1930–1940," *Isis*, vol. 77, 1986, pp. 450–472.
5. Seymour S. Cohen, "Finally the Beginnings of Molecular Biology," *TIBS*, vol. 11, 1986, pp. 92–93.
6. Philip J. Pauly, *Controlling Life: Jacques Loeb and the Engineering Ideal in Biology*, Oxford University Press, Oxford, 1987.
7. Wendell M. Stanley, "Isolation of a Crystalline Protein Possessing the Properties of Tobacco Mosaic Virus," *Science*, vol. 81, 1935, pp. 644–645.
8. Lily E. Kay, "The Twilight Zone of Life: The Crystallization of the Tobacco Mosaic Virus," 7th Course of the International School of the History of Biological Sciences, Ischia, 19–28 June 1990.
9. Max A. Lauffer, "Contributions of Early Research on Tobacco Mosaic Virus," *TIBS*, vol. 9, 1984, pp. 369–371. In 1934, Max Schlesinger had shown that another phage was made of nucleoprotein: Max Schlesinger, "Zur Frage der Chemischen Zusammensetzung des Bakteriophagen," *Biochem. Z.*, vol. 273, 1934, pp. 306–311.
10. What Stanley thought was important was that his result agreed with an autocatalytic model of life, and of proteins, which are its key constituent. This model was widely accepted by the Rockefeller Institute biologists Northrop, Bergmann, and Niemann (see Chapter 12). It explained the self-activation of proteolytic enzymes, such as trypsin, as well as the replication of bacteriophages and of multienzyme complexes in protein synthesis (Olby, *The Path to the Double Helix*, chap. 9).
11. Kay, "The Twilight Zone of Life."

7. The Role of the Physicists

1. "Physicists" is the general term used to denote scientists studying a wide range of disciplines: mathematicians, theoretical physicists, specialists in mechanics, and so on.
2. Bernard T. Feld and Gertrud Weiss Szilard, *The Collected Papers of Leo Szilard: Scientific Papers,* MIT Press, Cambridge, Mass., 1972; William Lanouette and Bela Silard, *Genius in the Shadows: A Biography of Leo Szilard, the Man behind the Bomb,* Charles Scribner's Sons, New York, 1992. A list of physicists involved in the biological revolution and a discussion of their contributions can be found in a paper that already seems "historic": Donald Fleming, "Emigré Physicists and the Biological Revolution," in Donald Fleming and Bernard Bailyn (eds.), *The Intellectual Migration: Europe and America,* 1930–1960, The Belknap Press of Harvard University Press, Cambridge, Mass., 1969.
3. Horace F. Judson, *The Eighth Day of Creation: The Makers of the Revolution in Biology,* Simon and Schuster, New York, 1979, p. 228. Gamow was also a well-known popular science writer who created the character Mr. Tomkins.
4. Szilard and Gamow were among those émigré scientists in the United States who were suspected of having helped the USSR to acquire information on the atomic bomb. See Pavel A. Sudoplatov, Anatoli P. Sudoplatov, Jerrold L. Schecter, and Leona P. Schecter, *Special Tasks: The Memoirs of an Unwanted Witness, a Soviet Spymaster,* Little, Brown and Company, Boston, 1994.
5. Nicholas C. Mullins, "The Development of a Scientific Specialty: The Phage Group and the Origins of Molecular Biology," *Minerva,* vol. 10, 1972, pp. 51–82.
6. François Jacob, *La logique du vivant,* Gallimard, Paris, 1970 (translated as *The Logic of Life,* Pantheon Press, New York, 1974).
7. Thomas S. Kuhn, *The Structure of Scientific Revolutions,* University of Chicago Press, Chicago, 1970.
8. Andrew Hodges, *Alan Turing: The Enigma of Intelligence,* Burnett Books Limited, London, 1983; Philippe Breton, *Histoire de l'informatique,* La Découverte, Paris, 1987.
9. R. V. Jones, *The Wizard War: British Scientific Intelligence, 1939–1945,* Coward McCann and Geoghan, New York, 1978; F. H. Hinsley and Alan Stripp (eds.), *Codebreakers: The Inside Story of Bletchley Park,* Oxford University Press, Oxford, 1993.
10. F. Aaserud, *Redirecting Science: Niels Bohr, Philanthropy and the Rise of Nuclear Physics,* Cambridge University Press, Cambridge, England, 1990; Abraham Pais, *Niels Bohr's Times in Physics: Philosophy and Polity,* Clarendon Press, Oxford, 1991, pp. 388–394.

11. Niels Bohr, "Light and Life," *Nature*, vol. 131, 1933, pp. 421–423 and 457–459.
12. Lily E. Kay, "Quanta of Life: Atomic Physics and the Reincarnation of Phage," *Hist. Phil. Life Sci.*, vol. 14, 1992, pp. 3–21. Niels Bohr's ideas on complementarity evolved throughout his life: from a complementarity between reductionism and vitalism, he passed to a complementarity between mechanism and finalism. His last article on biology, "Light and Life Revisited" (1962), did not refer to the notion of complementarity (Pais, *Niels Bohr's Times in Physics*, pp. 441–444).
13. Erwin Schrödinger, *What Is Life?* Cambridge University Press, Cambridge, England, 1944; Robert C. Olby, "Schrödinger's Problem: What Is Life?" *J. Hist. Biol.*, vol. 4, 1971, pp. 119–148. Schrödinger left Austria shortly after its annexation by Germany in 1938.
14. Gunther S. Stent, "That Was the Molecular Biology That Was," *Science*, vol. 160, 1968, pp. 390–395.
15. Erwin Schrödinger, *What Is Life?* (p. 21, Cambridge University Press, 1992). Schrödinger's position is a direct transposition of the mechanistic and deterministic view of Descartes, in terms of genes and chromosomes: "If all the parts of the seed of some particular animal, for example man, were known exactly, from this alone the face and the form of each of its limbs could be deduced, for entirely mathematical and precise reasons" *(Description du corps humain, Œuvres complètes*, ed. Adam-Tannery, vol. 11, p. 277).
16. Max F. Perutz, "Erwin Schrödinger's 'What Is Life?' and Molecular Biology," in *Schrödinger: Centenary Celebration of a Polymath*, C. V. Kilmister (ed.), Cambridge University Press, Cambridge, England, 1987.
17. Pnina Abir-Am, "Themes, Genres and Orders of Legitimation in the Consolidation of New Scientific Disciplines: Reconstructing the Historiography of Molecular Biology," *Hist. Sci.*, vol. 23, 1985, pp. 75–117.
18. Michel Morange, "Schrödinger et la biologie moléculaire," *Fundamenta Scientiae*, vol. 4, 1983, pp. 219–234.
19. Schrödinger, *What Is Life?* p. 21.
20. Edward Yoxen, "Where Does Schrödinger's 'What Is Life?' Belong in the History of Molecular Biology?" *Hist. Sci.*, vol. 17, 1979, pp. 17–52.
21. Ibid.
22. Walter Moore, *Schrödinger: Life and Thought*, Cambridge University Press, Cambridge, England, 1989.
23. Erwin Schrödinger, *My View of the World*, Part I: *Seek for the Road*, Cambridge University Press, Cambridge, England, 1964.
24. Morange, "Schrödinger et la biologie moléculaire."
25. William M. Johnston, *The Austrian Mind*, University of California Press, 1972.

8. The Influence of the Rockefeller Foundation

1. Robert E. Kohler, "The Management of Science: The Experience of Warren Weaver and the Rockefeller Foundation Programme in Molecular Biology," *Minerva,* vol. 14, 1976, pp. 279–306; Pnina Abir-Am, "The Discourse of Physical Power and Biological Knowledge in the 1930s: A Reappraisal of the Rockefeller Foundation's 'Policy' in Molecular Biology," *Soc. Stud. Sci.,* vol. 12, 1982, pp. 341–382; John A. Fuerst, Ditta Bartels, Robert Olby, Edward J. Yoxen, and Pnina Abir-Am, "Responses and Replies to P. Abir-Am: Final Response of P. Abir-Am," *Soc. Stud. Sci.,* vol. 14, 1984, pp. 225–263; Lily E. Kay, *The Molecular Vision of Life: Caltech, the Rockefeller Foundation and the Rise of the New Biology,* Oxford University Press, Oxford, 1993. Other foundations also played an important role in the birth of molecular biology, such as those that financed research on viruses (see Chapters 6 and 15).
2. Robert E. Kohler, *Partners in Science: Foundations and Natural Scientists, 1900–1945,* University of Chicago Press, Chicago, 1991.
3. Kohler, "The Management of Science," p. 290.
4. Ibid., p. 291.
5. Warren Weaver, "Molecular Biology: Origins of the Term," *Science,* vol. 170, pp. 591–592.
6. Doris T. Zallen, "The Rockefeller Foundation and French Research," *Les Cahiers pour l'histoire du CNRS,* vol. 5, 1989, pp. 35–58.
7. Jean-François Picard, *La République des savants. La recherche française et le CNRS,* Flammarion, Paris, 1990.
8. Abir-Am, "The Discourse of Physical Power."
9. Ibid., pp. 361–367; Pnina Abir-Am, "The Assessment of Interdisciplinary Research in the 1930s: The Rockefeller Foundation and Physico-Chemical Morphology," *Minerva,* vol. 26, 1988, pp. 153–176.
10. All of them were founder members of a theoretical biology club: Pnina G. Abir-Am, "The Biotheoretical Gathering, Trans-Disciplinary Authority and the Incipient Legitimation of Molecular Biology in the 1930s: New Perspective on the Historical Sociology of Science," *Hist. Sci.,* vol. 25, 1987, pp. 1–70; Pnina G. Abir-Am and Dorinda Outram (eds.), *Uneasy Careers and Intimate Lives: Women in Science,* 1789–1979, Rutgers University Press, New Brunswick, 1987. Chapter 12 is devoted to Dorothy Wrinch. Joseph Needham later turned to the study of Chinese science. Information on this early period of his life can be found in Joseph Needham, *Un taoïste d'honneur: autobiographie. De l'embryologie à la civilisation chinoise. Entretiens avec Didier Gazagnadou,* Éd. du Félin, Paris, 1991. On J. H. Woodger and his critique of methodological reductionism, see Nils Roll-Hansen, "E. S. Russell and J. H. Woodger: The Failure of Two

Twentieth-Century Opponents of Mechanistic Biology," *J. Hist. Biol.,* vol. 17, 1984, pp. 399–428; Rose-Marie Ponsot, "C. H. Waddington, l'évolution d'un évolutionniste," unpublished doctoral thesis, Université Paris-I, Paris, 1987.

11. Tim J. Horder and Paul Weindling, "Hans Spemann and the Organizer," pp. 183–242, in Tim Horder, Jan Witkowski, and C. C. Wylie (eds), *A History of Embryology,* Cambridge University Press, Cambridge, England, 1986. For an overall survey of his research, see Jan Witkowski's very clear study "The Hunting of the Organizer: An Episode in Biochemical Embryology," *TIBS,* vol. 10, 1985, pp. 379–381; Scott F. Gilbert, "Induction and the Origins of Developmental Genetics," in Scott F. Gilbert (ed.), *Developmental Biology,* vol. 7, *A Conceptual History of Modern Embryology,* Plenum Press, New York, 1991, pp. 181–206.
12. This is Bloor's famous symmetry principle, referred to in the Introduction.
13. J. M. W. Slack, *From Egg to Embryo,* Cambridge University Press, Cambridge, England, 1983 (cited by Jan Witkowski, "The Hunting of the Organizer").
14. Abir-Am and Outram, *Uneasy Careers and Intimate Lives,* p. 267.
15. Similarly, the patrons of Caltech were not the creators of molecular biology, contrary to Kay's suggestion, even though the research groups from this institute perhaps contributed more than others. See Kay, *The Molecular Vision of Life.*
16. The Institut de Biologie Physico-Chimique of Paris, in which Ephrussi and Beadle worked, was created with objectives that were very similar to those that were later followed by the Rockefeller Foundation: Antoine Danchin, *Physique, chimie, biologie, un demi-siècle d'interactions: 1927–1977,* text written for the fiftieth anniversary of Institut de Biologie Physico-Chimique, Paris, 1977.
17. Robert Bud, *The Uses of Life: A History of Biotechnology,* Cambridge University Press, Cambridge, England, 1993.

9. A New World View

1. William Bulloch, *The History of Bacteriology,* Oxford University Press, 1938 (reprinted Dover Publications Inc., New York, 1979); Thomas D. Brock, *Robert Koch: A Life in Medicine and Bacteriology,* Science Tech. Publishers, Madison, 1988.
2. A. J. P. Martin and R. L. M. Synge, "A New Form of Chromatogram Employing Two Liquid Phases: 1. A Theory of Chromatography; 2. Application to the Micro-Determination of the Higher Monoamino-Acids in Proteins," *Biochem. J.,* vol. 35, 1941, pp. 1358–1368; A. H. Gordon, A. J. P. Martin, and R. L. M. Synge, "Partition Chromatography in the Study of Protein Constituents," *Biochem. J.,* vol. 37, 1943, pp. 79–86.

3. R. Consden, A. H. Gordon, and A. J. P. Martin, "Qualitative Analysis of Proteins: A Partition Chromatographic Method Using Paper," *Biochem. J.*, vol. 38, 1944, pp. 224–232. Paper chromatography had already been used by the German chemical industry (in particular the dye and oil industry) in the nineteenth century; this method had later been forgotten (reported by A. H. Gordon in P. R. Srinivasan, Joseph S. Fruton, and John T. Edsall, *The Origins of Modern Biochemistry: A Retrospect on Proteins,* New York Academy of Science, New York, 1979).
4. Erwin Chargaff, *Hericlatean Fire: Sketches from a Life before Nature,* Rockefeller University Press, New York, 1978, pp. 90–93.
5. Ion exchange resins, long used in organic chemistry, played an important role in the Manhattan Project in separating the products of nuclear fission (Arthur Kornberg, *For the Love of Enzymes: The Odyssey of a Biochemist,* Harvard University Press, Cambridge, Mass., 1989). The first resins of this kind to be used in molecular biology were polymethacrylic acid resins (Amberlite IRC 50) and sulfonated polystyrene resins. Many other resins (Dowex, DEAE-cellulose) appeared in the next few years, some of which were better suited to protein and polypeptide separation. Stanford Moore and William Stein used this technique to develop automatic amino acid analyzers. In 1948 Moore and Stein were also the first to use automatic fraction collectors, which simplified column chromatography considerably. See Stanley Moore and William H. Stein, "Chemical Structures of Pancreatic Ribonuclease and Deoxyribonuclease," *Science,* vol. 180, 1973, pp. 458–464.
6. Jerker Porath and P. Flodin, "Gel Filtration: A Method for Desalting and Group Separation," *Nature,* vol. 183, 1959, pp. 1657–1659.
7. Rolf Axen, Jerker Porath, and Sverker Ernback, "Chemical Coupling of Peptides and Proteins to Polysaccharides by Means of Cyanogen Halides," *Nature,* vol. 214, 1967, pp. 1302–1304; Pedro Cuatrecasas, Meir Wilchek, and Christian Anfinsen, "Selective Enzyme Purification by Affinity Chromatography," *Proc. Natl. Acad. Sci. USA,* vol. 61, 1968, pp. 636–643.
8. Robert Olby, *The Path to the Double Helix,* Macmillan, London, 1974, pp. 11–21 et seq; Lily E. Kay, *The Molecular Vision of Life: Caltech, the Rockefeller Foundation and the Rise of the New Biology,* Oxford University Press, Oxford, 1993, p. 112 et seq.; Boelie Elzen, "Two Ultracentrifuges: A Comparative Study of the Social Construction of Artefacts," *Soc. Stud. Sci.,* vol. 16, 1986, pp. 621–662.
9. Milton Kerker, "The Svedberg and Molecular Reality," *Isis,* vol. 67, 1976, pp. 190–216.
10. The Svedberg, "The Ultra-Centrifuge and the Study of High-Molecular Compounds," *Nature,* vol. 139, 1937, pp. 1051–1062.
11. Dorothy M. Wrinch, "The Pattern of Proteins," *Nature,* vol. 137, 1936, pp. 411–412.

12. Linus Pauling and C. Niemann, "The Structure of Proteins," *Science,* vol. 61, 1939, pp. 1860–1867.
13. Milton Kerker, "The Svedberg and Molecular Reality: An Autobiographical Postscript," *Isis,* vol. 77, 1986, pp. 278–282.
14. Lily E. Kay, "W. M. Stanley's Crystallization of the Tobacco Mosaic Virus, 1930–1940," *Isis,* vol. 77, 1986, pp. 450–472.
15. Albert Claude, "The Coming of Age of the Cell: The Inventory of Cells by Fractionation, Biochemistry and Electron Microscopy Has Affected Our Status and Thinking," *Science,* vol. 189, 1975, pp. 433–435.
16. Elzen, "Two Ultracentrifuges."
17. Lily E. Kay, "Laboratory Technology and Biological Knowledge: The Tiselius Electrophoresis Apparatus, 1930–1945," *Hist. Phil. Life Sci.,* vol. 10, 1988, pp. 51–72.
18. R. Consden, A. H. Gordon, and A. J. P. Martin, "Ionophoresis in Silica Jelly: A Method for the Separation of Amino-Acids and Peptides," *Biochem. J.,* vol. 40, 1946, pp. 33–41.
19. Samuel Raymond and Lewis Weintraub, "Acrylamide Gel as a Supporting Medium for Zone Electrophoresis," *Science,* vol. 130, 1959, p. 711.
20. Kay, "Laboratory Technology and Biological Knowledge," p. 51. Arne Tiselius made significant contributions to chromatography: see the review by A. J. P. Martin and R. L. M. Synge, "Analytical Chemistry of the Proteins," *Adv. Prot. Chem.,* vol. 2, 1945, pp. 1–83; and the autobiographical account of Arne Tiselius, "Reflections from Both Sides of the Counter," *Ann. Rev. Biochem,* vol. 37, 1968, pp. 1–24.
21. Ronald Bentley, "The Use of Stable Isotopes at Columbia University's College of Physicians and Surgeons," *TIBS,* vol. 10, 1985, pp. 171–174.
22. Rudolph Schonheimer, *The Dynamic State of Body Constituents,* Harvard University Press, Cambridge, Mass., 1942.
23. Jacques Monod, "From Enzymatic Adaptation to Allosteric Transitions," *Science,* vol. 154, 1966, p. 477.
24. Robert E. Kohler, Jr., "Rudolf Schonheimer, Isotopic Tracers and Biochemistry in the 1930s," *Historical Studies in the Physical Sciences,* vol. 8, 1977, pp. 257–298.
25. Doris T. Zallen, "The Rockefeller Foundation and Spectroscopy Research: The Programs at Chicago and Utrecht," *J. Hist. Biol.,* vol. 25, 1992, pp. 67–89.
26. In particular, in providing quantitative measures of nucleic acids, as in Chargaff's work.
27. V. E. Cosslet, "The Early Years of Electron Microscopy in Biology," *TIBS,* vol. 10, 1985, pp. 361–363; Nicolas Rasmussen, "Making a Machine Instrumental: RCA and the Wartime Origins of Biological Electron Microscopy in America, 1940–1945," *Stud. Hist. Phil. Sci.,* vol. 27, 1996, pp. 311–349.
28. For example, mesosomes. Nicolas Rasmussen, "Facts, Artifacts and Mesosomes: Practicing Epistemology with the Electron Microscope," *Stud. Hist.*

Phil. Sci., vol. 24, 1993, pp. 227–265. The limitations of electron microscopy are such that the work of specialists in this field is not representative of biological research.

10. The Role of Physics

1. Horace F. Judson, *The Eighth Day of Creation: The Makers of the Revolution in Biology,* Simon and Schuster, New York, 1979. See in particular pp. 606–607.
2. Arthur Kornberg, "Molecular Origins," *Nature,* vol. 214, 1967, p. 538.
3. Alain Prochiantz, "L'illusion physicaliste dans les sciences de la vie," *Revue Internationale de psychopathologie,* vol. 8, 1992, pp. 553–569. Even if that were the case, Schrödinger's role would still not have been negligible. Analogies and metaphors have an important place in scientific research.
4. Olga Amsterdamska, "Stabilizing Instability: The Controversy over Cyclogenic Theories of Bacterial Variation during the Interwar Period," *J. Hist. Biol.*, vol. 24, 1991, p. 221.
5. Donald Fleming and Bernard Bailyn (eds.), *The Intellectual Migration: Europe and America,* 1930–1960, the Belknap Press of Harvard University Press, Cambridge, Mass., 1969; Paul K. Hoch, "Migration and the Generation of New Scientific Ideas," *Minerva,* vol. 25, 1987, pp. 209–237. Molecular biology developed in an "international space." See Pnina Abir-Am, "From Multidisciplinary Collaboration to Transnational Objectivity: International Space as Constitutive of Molecular Biology, 1930–1970," in Elisabeth Crawford, Terry Shin, and Sverker Sorlin (eds), *Denationalizing Science: The Contexts of International Scientific Practice,* Kluwer Academic Publishers, Dordrecht, 1993, pp. 153–186.
6. Evelyn Fox Keller, "Physics and the Emergence of Molecular Biology: A History of Cognitive and Political Synergy," *J. Hist. Biol.*, vol. 23, 1990, pp. 389–409.
7. "Everything suggests that the logic of heredity can be compared to that of a calculator. Seldom has the model of an epoch found such a faithful application." François Jacob, *La logique du vivant,* Gallimard, Paris, 1970 (translated as *The Logic of Life,* Pantheon Press, New York, 1974). The chronologies are in fact remarkably parallel—1936: the Turing machine, first "theoretical" concept of a computer; 1944: Avery's experiment; 1945: conception of the first computer (EDVAC) by John von Neumann; 1948: creation of cybernetics by Norbert Wiener, publication of the theory of information by Claude Shannon; 1953: discovery of the double helix structure of DNA and early reflections on the genetic code. See Norbert Wiener, "Cybernetics, or Control and Communication in the Animal and the Machine," *Actualités Sci. Ind.*, no. 1053, Hermann (Paris), MIT Press (Cambridge, Mass.), and Wiley (New

York), 1948; Claude E. Shannon, "A Mathematical Theory of Communication," *Bell System Technical Journal,* vol. 27, 1948, pp. 379–423 and 623–656. In the years after the Second World War, John von Neumann made many contacts with the founders of molecular biology (William Aspray, *John von Neumann and the Origins of Modern Computing,* MIT Press, Cambridge, Mass., 1990, p. 181 et seq.).

8. Andrew Hodges, *Alan Turing: The Enigma of Intelligence,* Burnett Books Limited, London, 1983. Alan Turing was himself interested in biology. One of his last publications was devoted to morphogenesis: A. M. Turing, "The Chemical Basis of Morphogenesis," *Ph. Trans. Roy. Soc.,* London, series B, vol. 237, 1952, pp. 37–72. The relations between cyberneticists and molecular biologists, such as Delbrück, were at first difficult: Steve J. Heims, *Constructing a Social Science for Postwar America: The Cybernetics Group,* 1946–1953, MIT Press, Cambridge, Mass., 1993, pp. 93–96. Although at the end of the 1940s and at the beginning of the 1950s information theory did not have a practical influence on molecular biology, it did survive on level of discourse and became widely used at the beginning of the 1960s: Lily E. Kay, "Who Wrote the Book of Life? Information and the Transformation of Molecular Biology, 1945–55," *Science in Context,* vol. 8, 1995, pp. 609–634.

11. The Discovery of the Double Helix

1. Watson has written a lively account of this period; see James D. Watson, *The Double Helix: A Personal Account of the Discovery of the Structure of DNA,* Weidenfeld and Nicholson, London, 1968; rev. ed. by Gunther S. Stent, Weidenfeld and Nicholson, London, 1981. The British historian Robert Olby has written *The Path to the Double Helix,* Macmillan, London, 1974. Horace F. Judson has reconstituted the periods before and after this discovery, using a large number of interviews; see *The Eighth Day of Creation: The Makers of the Revolution in Biology,* Simon and Schuster, New York, 1979. These are the main sources for the material to be found in this chapter.
2. Francis Crick, *What Mad Pursuit: A Personal View of Scientific Discovery,* Basic Books, New York, 1988.
3. On the history of the Cambridge molecular biology laboratory, see Soraya De Chadarevian, "The Architecture of Proteins: Building the Laboratory of Molecular Biology at Cambridge/UK," In *Experimentalsysteme in den biologisch-medizinischen Wissenschaften: Objekte, Differenzen Konjunkturen,* M. Hagner and H.-J. Rheinberger (eds.), Akademie Verlag, Berlin, 1994, pp. 181–200.

4. John M. Thomas and David Phillips (eds.), *Selections and Reflections: The Legacy of Sir Lawrence Bragg,* Science Reviews, London, 1991.
5. Crick, *What Mad Pursuit.*
6. Anne Sayre, *Rosalind Franklin and DNA,* W. W. Norton and Co., New York, 1975.
7. Linus Pauling and Robert B. Corey, "Two Hydrogen-Bonded Spiral Configurations of the Polypeptide Chain," *J. Am. Chem. Soc.,* vol. 72, 1950, p. 5349; Linus Pauling, Robert B. Corey, and H. R. Branson, "Two Hydrogen-Bonded Helical Configurations of the Polypeptide Chain," *Proc. Natl. Acad. Sci. USA,* vol. 37, 1951, pp. 205–211; Linus Pauling and Robert B. Corey, seven successive papers in *Proc. Natl. Acad. Sci. USA,* vol. 37, 1951, pp. 235–285.
8. Olby, *The Path to the Double Helix,* chap. 4.
9. Sir Lawrence Bragg, John C. Kendrew, and Max F. Perutz, "Polypeptide Chain Configuration in Crystalline Proteins," *Proc. Roy. Soc. A,* vol. 203, 1950, pp. 321–357.
10. Alexander Rich and Norman Davidson, *Structural Chemistry and Molecular Biology,* W. H. Freeman and Co., San Francisco, 1968.
11. Linus Pauling and Robert B. Corey, "A Proposed Structure for the Nucleic Acids," *Proc. Natl. Acad. Sci. USA,* vol. 39, 1953, pp. 84–97.
12. Keith L. Manchester, "Did a Tragic Accident Delay the Discovery of the Double Helical Structure of DNA?" *TIBS,* vol. 20, 1995, pp. 126–128.
13. James D. Watson and Francis H. C. Crick, "A Structure for Deoxyribose Nucleic Acid," *Nature,* vol. 171, 1953, pp. 737–738; Maurice H. F. Wilkins, Alexander R. Stokes, and H. R. Wilson, "Molecular Structure of Deoxy Pentose Nucleic Acid," *Nature,* vol. 171, 1953, pp. 738–740; Rosalind E. Franklin and Raymond G. Gosling, "Molecular Configuration in Sodium Thymonucleate," *Nature,* vol. 171, 1953, pp. 740–741.
14. James D. Watson and Francis H. C. Crick, "Genetical Implications of the Structure of Deoxyribonucleic Acid," *Nature,* vol. 171, 1953, pp. 964–967.
15. John D. Bernal, "Structure of Proteins," *Nature,* vol. 143, 1939, pp. 663–667.
16. Ernest P. Fischer and Carol Lipson, *Thinking about Science: Max Delbrück and the Origins of Molecular Biology,* W. W. Norton, New York, 1988.
17. The concept of tinkering *(bricolage)* was introduced in 1977 by François Jacob to describe evolution's action on life (see Chapter 17). See François Jacob, "Evolution and Tinkering," *Science,* vol. 196, 1977, pp. 1161–1166. As Jacob points out in his article, this concept also applies to scientific research. An interpretation slightly different from that presented here of the usefulness of this idea in describing the behavior of scientists can be found in Karin D. Knorr-Cetina, *The Manufacture of Knowledge: An Essay on the Constructivist and Contextual Nature of Science,* Pergamon Press, Oxford, 1981, pp. 34–35.

18. This was how, in 1954, Watson and Crick tried to reconstruct rationally the discovery of the double helix structure of DNA. Francis H. C. Crick and James D. Watson, "The Complementary Structure of Deoxyribonucleic Acid," *Proc. Roy. Soc. Ser. A,* vol. 223, 1954, pp. 80–96.
19. Gunther S. Stent, "Prematurity and Uniqueness in Scientific Discovery," *Scientific American,* vol. 227, December 1972, pp. 84–93.
20. Crick, *What Mad Pursuit.*
21. "Genes, like cells, organisms, or species, are good candidates for the status of 'natural things'—far more so than the 'particles' of contemporary physics. The photon, the muon, and the quark are relatively straightforward when described in terms of their lawful behavior; however, their status as 'things' is more problematic. The history of the science of heredity has tended to dissociate the conceptual universes of order and of measurement, which have been too readily identified by a tradition of the philosophy of science centered on a few chapters in the history of mathematical physics." Jean Gayon, "De la mesure à l'ordre: histoire philosophique du concept d'hérédité," in *Passion des Formes,* ENS Fontenay-St Cloud Editions, Paris, 1994.
22. Sayre, *Rosalind Franklin and DNA.*
23. Rosalind Franklin had made some important steps toward the correct structure, which explains why she immediately accepted Watson and Crick's results. See A. Klug, "Rosalind Franklin and the Discovery of the Structure of DNA," *Nature,* vol. 219, 1968, pp. 808–810, 843–844.
24. Bernadette Bensaude-Vincent, "Une mythologie révolutionnaire dans la chimie française," *Annals of Science,* vol. 40, 1983, pp. 189–196.
25. Pnina Abir-Am, "From Biochemistry to Molecular Biology: DNA and the Acculturated Journey of the Critic of Science Erwin Chargaff," *Hist. Phil. Life Sci.,* vol. 2, 1980, pp. 3–60; Pnina Abir-Am, "How Scientists View Their Heroes: Some Remarks on the Mechanism of Myth Construction," *J. Hist. Biol.,* vol. 15, 1982, pp. 281–315.
26. To borrow the title of Robert Olby's book, *The Path to the Double Helix.*
27. Crick, *What Mad Pursuit.*
28. Monica Winstanley, "Assimilation into the Literature of a Critical Advance in Molecular Biology," *Soc. Stud. Sci.,* vol. 6, 1976, pp. 545–549; Barak Gaster, "Assimilation of Scientific Change: The Introduction of Molecular Genetics into Biology Textbooks," *Soc. Stud. Sci.,* vol. 20, 1990, pp. 431–454.
29. Judson, *The Eighth Day of Creation,* p. 261 et seq.
30. Max Delbrück and Gunther S. Stent, "On the Mechanism of DNA Replication," in W. D. McElroy and B. Glass (eds.), *The Chemical Basis of Heredity,* Johns Hopkins Press, Baltimore, 1957, pp. 699–736.
31. John B. S. Haldane, *New Paths in Genetics,* Allen and Unwin, London, 1941, p. 44.

32. Judson, *The Eighth Day of Creation,* p. 188 et seq.
33. Matthew Meselson and Franklin W. Stahl, "The Replication of DNA in *Escherichia coli,*" *Proc. Natl. Acad. Sci. USA,* vol. 44, 1958, pp. 671–682.

12. Deciphering the Genetic Code

1. James D. Watson and Francis H. C. Crick, "Genetical Implications of the Structure of Deoxyribonucleic Acid," *Nature,* vol. 171, 1953, pp. 964–967.
2. Horace F. Judson, *The Eighth Day of Creation: The Makers of the Revolution in Biology,* Simon and Schuster, New York, 1979, p. 261 et seq.
3. By an "unlucky" chance the distance separating amino acids in an extended protein chain is 3.3–3.4 Å (this had been noted by William Astbury as early as 1938). This value is identical to that separating two successive nucleotides in a DNA molecule. This fact led to a number of hypotheses. It also explains why a hypothesis analogous to Gamow's had previously been proposed by Pauling and Corey when they published their erroneous model of the structure of DNA. See Linus Pauling and Robert B. Corey, "A Proposed Structure for the Nucleic Acids," *Proc. Natl. Acad. Sci. USA,* vol. 39, 1953, p. 96.
4. Francis H. C. Crick, John S. Griffith, and Leslie E. Orgel, "Codes without Commas," *Proc. Natl. Acad. Sci. USA,* vol. 43, 1957, pp. 416–421.
5. Carl R. Woese, *The Genetic Code: The Molecular Basis for Genetic Expression,* Harper and Row, New York, 1967, chap. 2; Martynas Ycas, *The Biological Code,* North Holland Publishing Company, Amsterdam, 1969; Jan A. Witkowski, "The 'Magic' of Numbers," *TIBS,* vol. 10, 1985, pp. 139–141.
6. Francis H. C. Crick, "The Present Position of the Coding Problem," *Brookhaven Symposia,* vol. 12, 1959, pp. 35–39.
7. Max Bergmann and Carl Niemann, "Newer Biological Aspects of Protein Chemistry," *Science,* vol. 86, 1937, pp. 187–190.
8. Archer J. P. Martin and Richard L. M. Synge, "A New Form of Chromatogram Employing Two Liquid Phases: 1. A Theory of Chromatography; 2. Application to the Micro-Determination of the Higher Monoamino-Acids in Proteins," *Biochem. J.,* vol. 35, 1941, pp. 1358–1361; A. H. Gordon, Archer J. P. Martin, and Richard L. M. Synge, "Partition Chromatography in the Study of Protein Constituents," *Biochem. J.,* vol. 37, 1943, pp. 79–86; R. Consden, A. H. Gordon, and Archer J. P. Martin, "Qualitative Analysis of Proteins: A Partition Chromatographic Method Using Paper," *Biochem. J.,* vol. 38, 1944, pp. 224–232.
9. Frederick Sanger, "Sequences, Sequences and Sequences," *Ann. Rev. Biochem.,* vol. 57, 1988, pp. 1–28. Sanger profited from the fact that proteases had

become relatively well understood, in particular as a result of the work of Bergmann's group. To carry out this work, he developed a reagent that was specific to the N-terminal part of amino acids, subsequently known as Sanger's reagent. On Sanger's relations with the Cambridge molecular biologists, see Soraya de Chadarevian, "Sequences, Conformation, Information: Biochemists and Molecular Biologists in the 1950s," *J. Hist. Biol.*, vol. 29, 1996, pp. 361–386.

10. A. P. Ryle, Frederick Sanger, L. F. Smith, and Ruth Kital, "The Disulphide Bonds of Insulin," *Biochem. J.*, vol. 60, 1955, pp. 541–556; Frederick Sanger, "Chemistry of Insulin: Determination of the Structure of Insulin Opens the Way to Greater Understanding of Life Processes," *Science*, vol. 129, 1959, pp. 1340–1344. The structure of a small antibiotic polypeptide, gramicidin, had been determined slightly earlier, along with—thanks to the work of du Vigneaud—the sequence of two small pituitary gland hormones, ocytocin and vasopressin.
11. Linus Pauling, Harvey A. Itano, Seymour J. Singer, and Ibert C. Wells, "Sickle Cell Anemia, a Molecular Disease," *Science*, vol. 110, 1949, pp. 543–548.
12. James V. Neel, "The Inheritance of Sickle Cell Anemia," *Science*, vol. 110, 1949, pp. 64–66.
13. Vernon M. Ingram, "A Specific Chemical Difference between the Globins of Normal Human and Sickle-Cell Anaemia Haemoglobin," *Nature*, vol. 178, 1956, pp. 792–794; Vernon M. Ingram, "Gene Mutations in Human Haemoglobin: The Chemical Difference between Normal and Sickle Cell Haemoglobin," *Nature*, vol. 180, 1957, pp. 326–328.
14. Joseph S. Fruton, "Proteolytic Enzymes as Specific Agents in the Formation and Breakdown of Proteins," *Cold Spring Harbor Symp. Quant. Biol*, vol. 9, 1941, pp. 211–217; Ditta Bartels, "The Multi-Enzyme Programme of Protein Synthesis: Its Neglect in the History of Biochemistry and Its Current Role in Biotechnology," *Hist. Phil. Life Sci.*, vol. 5, 1983, pp. 187–219.
15. Judson, *The Eighth Day of Creation*, p. 247.
16. The earliest data were obtained by Joachim Hammerling in 1934 on a giant unicellular alga, *Acetabularia mediterranea:* Joachim Hammerling, "Über formbildende Substanzen bei Acetabularia mediterranea, Räumliche und zeitliche Verteilung und ihre Herkunft," *Wilh. Roux. Arch. Entwicklungsmech. Org.*, vol. 131, 1934, pp. 1–81.
17. Torbjörn Caspersson and Jack Schultz, "Pentose Nucleotides in the Cytoplasm of Growing Tissues," *Nature*, vol. 143, 1939, pp. 602–603; Jean Brachet, "La localisation des acides pentose nucléiques dans les tissus animaux et les œufs d'amphibiens en voie de développement," *Archives de biologie*, vol. 53, 1942, pp. 207–257.
18. Alexander L. Dounce, "Duplicating Mechanism for Peptide Chain and Nucleic Acid Synthesis," *Enzymologia*, vol. 15, 1952, pp. 251–258. The term "template" was

introduced into biochemistry in 1904 by H. E. Armstrong to denote the "surface" on which enzymatic catalysis takes place (H. E. Armstrong, "Enzyme Action as Bearing on the Validity of the Ionic-Dissociation Hypothesis and on the Phenomena of Vital Change," *J. Chem. Soc.*, vol. 73, 1904, p. 537). The term thus suggested stereospecific recognition (see Chapter 1).

19. Alexander Dounce, "Nucleic and Template Hypotheses," *Nature*, vol. 172, 1953, pp. 541–542. This model was again supported by the similarity of the distance both between amino acids in an unfolded polypeptide chain and between nucleotides in the DNA molecule. (See note 3 above.)
20. André Boivin and Roger Vendrely, "Sur le rôle possible des deux acides nucléiques dans la cellule vivante," *Experientia*, vol. 3, 1947, pp. 32–34.
21. Linus Pauling, "A Theory of the Structure and Process of Formation of Antibodies," *J. Am. Chem. Soc.*, vol. 62, 1940, pp. 2643–2657.
22. For a more complete presentation of theories of antibody biosynthesis, see Ann-Marie Moulin, *Le Dernier Langage de la médecine: histoire de l'immunologie de Pasteur au SIDA*, PUF, Paris, 1991.
23. Karl Landsteiner, *The Specificity of Serological Reactions*, Charles C. Thomas, Springfield, Ill., 1936.
24. Linus Pauling, "A Theory of the Structure and Process of Formation of Antibodies," *J. Am. Chem. Soc.*, vol. 62, 1940, pp. 2643–2657.
25. F. Breinl and Felix Haurowitz, "Chemische Untersuchung des Präzipitates aus Hämoglobin und Anti-Hämoglobin Serum und Bemerkungen über die Natur der Antikörper," *Hoppe-Seyler's Zeitschrift für physiologische chemie*, vol. 192, 1930, pp. 45–57; Jerome Alexander, *Protoplasma*, vol. 14, 1931, p. 296; Stuart Mudd, "A Hypothetical Mechanism of Antibody Formation," *J. Immunol.*, vol. 23, 1932, pp. 423–427.
26. According to Karl Popper, what distinguishes a scientific theory from a nonscientific one is that the former can be refuted or falsified by experimentation. See Karl Popper, *The Logic of Scientific Discovery*, Hutchinson, London, 1959.
27. Experiments reported by Burnet: F. M. Burnet and F. Fenner, *The Production of Antibodies*, Macmillan, Melbourne, 1949.
28. Linus Pauling and Dan H. Campbell, "The Production of Antibodies *in vitro*," *Science*, vol. 95, 1942, pp. 440–441; Linus Pauling and Dan H. Campbell, "The Manufacture of Antibodies *in vitro*," *J. Exp. Med.*, vol. 76, 1942, pp. 211–220.
29. Felix Haurowitz, Paula Schwerin, and Saide Tunç, "The Mutual Precipitation of Proteins and Azoproteins," *Arch. Biochem.*, vol. 11, 1946, pp. 515–520.
30. Lily E. Kay, "Molecular Biology and Pauling's Immunochemistry: A Neglected Dimension," *Hist. Phil. Life Sci.*, vol. 11, 1989, pp. 211–219; Lily E. Kay, *The Molecular Vision of Life: Caltech, the Rockefeller Foundation and the Rise of the New Biology*, Oxford University Press, Oxford, 1993, chap. 6.
31. See, for example, MacFarlane Burnet, *The Clonal Selection Theory of Acquired Immunity*, Vanderbilt University Press, Nashville, 1959; Niels K. Jerne, "The

Natural Selection Theory of Antibody Formation," *Proc. Natl. Acad. Sci. USA*, vol. 41, 1955, pp. 849–857.

32. Linus Pauling, "Molecular Basis of Biological Specificity," *Nature*, vol. 248, 1974, pp. 769–771.
33. Felix Haurowitz, *Chemistry and Biology of Proteins*, Academic Press, New York, 1950; "The Mechanism of the Immunological Response," *Biol. Rev.*, vol. 27, 1952, pp. 247–280.
34. Jacques Monod, "The Phenomenon of Enzymatic Adaptation and Its Bearings on Problems of Genetics and Cellular Differentiation," *Growth Symposium*, vol. 11, 1947, pp. 223–289.
35. Jacques Monod and Melvin Cohn, "La biosynthèse induite des enzymes (adaptation enzymatique)," *Advances in Enzymology*, vol. 13, 1952, pp. 67–119.
36. Kay, "Molecular Biology and Pauling's Immunochemistry," pp. 211–219.
37. George W. Gray, "Pauling and Beadle," *Sci. Am.*, vol. 180, 1949, pp. 16–21, quotation pp. 19–20; mentioned by Kay, "Molecular Biology and Pauling's Immunochemistry."
38. Alfred H. Sturtevant, "Can Specific Mutations Be Induced by Serological Methods?" *Proc. Natl. Acad. Sci. USA*, vol. 30, 1944, pp. 176–178.
39. Moulin, *Le Dernier Langage de la médecine*, p. 176.
40. Sterling Emerson, "The Induction of Mutations by Antibodies," *Proc. Natl. Acad. Sci. USA*, vol. 30, 1944, pp. 179–183.
41. Seymour S. Cohen, "The Biochemical Origins of Molecular Biology," *TIBS*, vol. 9, 1984, pp. 334–336; Robert C. Olby, "Biochemical Origins of Molecular Biology: A Discussion," *TIBS*, vol. 11, 1986, pp. 303–305.
42. John Cairns, Gunther S. Stent, and James D. Watson (eds.), *Phage and the Origins of Molecular Biology*, Cold Spring Harbor Laboratory Press, Cold Spring Harbor, 1966, expanded ed., 1992; John C. Kendrew, "How Molecular Biology Started," *Sci. Am.*, vol. 216, 1967, pp. 141–144 (this idea was taken up and developed by Gunther S. Stent, "That Was the Molecular Biology That Was," *Science*, vol. 160, 1968, pp. 390–395).
43. Paul Zamecnik, "The Machinery of Protein Synthesis," *TIBS*, vol. 9, 1984, pp. 464–466; "Historical Aspects of Protein Synthesis," in P. R. Srinivasan, Joseph S. Fruton, and John T. Edsall, *The Origins of Modern Biochemistry: A Retrospect on Proteins*, the New York Academy of Science, New York, 1979, pp. 269–301; Hans-Jörg Rheinberger, "Experiment, Difference and Writing: I. Tracing Protein Synthesis," *Stud. Hist. Phil. Sci.*, vol. 23, 1992, pp. 305–331, and II. "The Laboratory Production of Transfer RNA," *Stud. Hist. Phil. Sci.*, vol. 23, 1992, pp. 389–422; Hans-Jörg Rheinberger, "Experiment and Orientation: Early Systems of in Vitro Protein Synthesis," *J. Hist. Biol.*, vol. 26, 1993, pp. 443–471. The very detailed studies by Hans-Jörg Rheinberger clearly show how the strategies chosen by scientists are constrained and directed by exper-

imental "resistance." For example, one of the most difficult steps was the development of methods for rupturing cells that did not lead to the inactivation of cellular proteins.

44. Mahlon Hoagland, *Towards the Habit of Truth: A Life in Science,* W. W. Norton, New York, 1990.
45. Judson, *The Eighth Day of Creation,* chap. 6.
46. Marvin R. Lamborg and Paul C. Zamecnik, "Amino Acid Incorporation into Proteins by Extracts of *E. coli,*" *Biochem. Biophys. Acta,* vol. 42, 1960, pp. 206–211.
47. Alfred Tissières, David Schlessinger, and François Gros, "Amino Acid Incorporation into Proteins by *Escherichia coli* ribosomes," *Proc. Natl. Acad. Sci. USA,* vol. 46, 1960, pp. 1450–1463.
48. These experiments are well described in Judson, *The Eighth Day of Creation,* pp. 470–489.
49. Johann H. Matthaei and Marshall W. Nirenberg, "Characteristics and Stabilization of DNAase Sensitive Protein Synthesis in *E. coli* Extracts," *Proc. Natl. Acad. Sci. USA,* vol. 47, 1961, pp. 1580–1588.
50. Marshall W. Nirenberg and Johann H. Matthaei, "The Dependence of Cell-Free Protein Synthesis in *E. coli* upon Naturally Occuring or Synthetic Polyribonucleotides," *Proc. Natl. Acad. Sci. USA,* vol. 47, 1961, pp. 1588–1602.
51. Judson, *The Eighth Day of Creation,* pp. 478 and 482; Robert C. Olby, "And on the Eighth Day. . . ," *TIBS,* vol. 4, 1979, pp. N215–N216.
52. Francis H. Crick, "The Recent Excitement in the Coding Problem," *Progress in Nucleic Acid Research,* Academic Press, New York, vol. 1, 1963, pp. 163–217.
53. Har G. Khorana, "Polynucleotide Synthesis and the Genetic Code," *Harvey Lectures* 1966–1967, series 62, Academic Press, New York, 1968, pp. 79–105.
54. Marshall W. Nirenberg and Philip Leder, "RNA Codewords and Protein Synthesis: The Effect of Trinucleotides upon the Binding of sRNA to Ribosomes," *Science,* vol. 145, 1964, pp. 1399–1407.
55. Francis H. C. Crick, Leslie Barnett, Sydney Brenner, and R. J. Watts-Tobin, "General Nature of the Genetic Code for Proteins," *Nature,* vol. 192, 1961, pp. 1227–1232.
56. Francis Crick, *What Mad Pursuit: A Personal View of Scientific Discovery,* Basic Books, New York, 1988.

13. The Discovery of Messenger RNA

1. Francis Crick, *What Mad Pursuit: A Personal View of Scientific Discovery,* Basic Books, New York, 1988.
2. Francis H. Crick, "On Protein Synthesis," *Symp. Soc. Exptl. Biol.,* vol. 12, pp. 138–163. The spontaneous folding hypothesis was based on a large number

of experiments on protein denaturation-renaturation that were begun in the 1930s and brilliantly developed on ribonuclease by Christian Anfinsen in the late 1950s. The results of these experiments, however, were not always clear.

3. Johann H. Matthaei and Marshall W. Nirenberg, "Characteristics and Stabilization of DNAase Sensitive Protein Synthesis in *E. coli* Extracts," *Proc. Natl. Acad. Sci. USA*, vol. 47, 1961, pp. 1580–1588.
4. Torbjörn Caspersson and Jack Schultz, "Pentose Nucleotides in the Cytoplasm of Growing Tissues," *Nature*, vol. 143, 1939, pp. 602–603; Torbjörn Caspersson, "Nukleinsaureketten und Genvermehrung," *Chromosoma*, vol. 1, 1940, pp. 605–619.
5. Jean Brachet, "La localisation des acides pentosenucléiques dans les tissus animaux et les œufs d'amphibiens en voie de développement," *Archives de biologie*, vol. 53, 1942, pp. 207–257.
6. Jean Brachet, "Recherches sur les interactions biochimiques entre le noyau et le cytoplasme chez les organismes unicellulaires. I. *Amoeba proteus*," *Biochem. Biophys. Acta*, vol. 18, 1955, pp. 247–268. For a re-evaluation of the role of the "Belgian" molecular biology group at Rouge-Cloître near Brussels, see Denis Thieffry and Richard M. Burian, "Jean Brachet's alternative scheme for protein synthesis," *TIBS*, vol. 21, 1996, pp. 114–117.
7. Ilana Löwy, "Variances in Meaning in Discovery Accounts: The Case of Contemporary Biology," *HSPS*, vol. 21, 1990, pp. 87–121.
8. This work was described by Palade in 1974, in his Nobel lecture: George Palade, "Intracellular Aspects of the Process of Protein Synthesis," *Science*, vol. 189, 1975, pp. 347–358. See also Hans-Jörg Rheinberger, "From Microsomes to Ribosomes: 'Strategies' of 'Representation,'" *J. Hist. Biol.*, vol. 28, 1995, pp. 49–89.
9. Heinz Fraenkel-Conrat, "The Role of the Nucleic Acid in the Reconstitution of Active Tobacco Mosaic Virus," *J. Am. Chem. Soc.*, vol. 78, 1956, pp. 882–883.
10. Arthur B. Pardee, François Jacob, and Jacques Monod, "The Genetic Control and Cytoplasmic Expression of 'Inducibility' in the Synthesis of β-galactosidase by *E. coli*," *J. Mol. Biol.*, vol. 1, 1959, pp. 165–178.
11. Monica Riley, Arthur Pardee, François Jacob, and Jacques Monod, "On the Expression of a Structural Gene," *J. Mol. Biol.*, vol. 2, 1960, pp. 216–225.
12. François Jacob, *La Statue Intérieure*, Odile Jacob, Paris, 1986, chap. 7 (translated as *The Statue Within: An Autobiography*, Basic Books, New York, 1988.
13. François Gros, *Les secrets du gène*, Odile Jacob, Paris, 1986, pp. 126–127.
14. Horace F. Judson, *The Eighth Day of Creation: The Makers of the Revolution in Biology*, Simon and Schuster, New York, 1979, pp. 428–436; Jacob, *La Statue Intérieure*, pp. 347–349; Crick, *What Mad Pursuit*.
15. Elliot Volkin and L. Astrachan, "Phosphorus Incorporation in *Escherichia coli* Ribonucleic Acid after Infection with Bacteriophage T2," *Virology*, vol. 2, 1956, pp. 149–161.

16. Jacob, *La Statue Intérieure,* pp. 415–427; Sydney Brenner, François Jacob, and Matthew Meselson, "An Unstable Intermediate Carrying Information from Genes to Ribosomes for Protein Synthesis," *Nature,* vol. 190, 1961, pp. 576–581.
17. François Gros, H. Hiatt, Walter Gilbert, Chuck G. Kurland, R. W. Risebrough, and James D. Watson, "Unstable Ribonucleic Acid Revealed by Pulse Labelling of *Escherichia coli,*" *Nature,* vol. 190, 1961, pp. 581–585; Gros, *Les Secrets du gène,* pp. 131–135.
18. Crick, *What Mad Pursuit.* Another example of a postmature discovery was discussed in Chapter 2—Beadle and Tatum's one gene–one enzyme hypothesis.
19. Andrei N. Belozersky and A. S. Spirin, "A Correlation between the Composition of Deoxyribonucleic Acid and Ribonucleic Acids," *Nature,* vol. 182, 1958, pp. 111–112.
20. There were several competing theories of cytoplasmic inheritance. The plasmon theory was opposed to the plasmagene theory and rejected the corpuscular nature of cytoplasmic inheritance.
21. The only study devoted to these theories is that of Jan Sapp, *Beyond the Gene: Cytoplasmic Inheritance and the Struggle for Authority in Genetics,* Oxford University Press, New York, 1987. See also Jan Sapp, "Hérédité Cytoplasmique et histoire de la génétique," in Jean-Louis Fischer and William H. Schneider, *Histoire de la génétique, pratique, techniques et théories,* ARPEM et Sciences en situation, Paris, 1990, pp. 231–246; Jan Sapp, "The Struggle for Authority in the Field of Heredity, 1900–1932: New Perspectives on the Rise of Genetics," *J. Hist. Biol.,* vol. 16, 1983, pp. 311–342; Jonathan Harwood, "Genetics and the Evolutionary Synthesis in Interwar Germany," *Annals of Science,* vol. 42, 1985, pp. 279–301, and *Styles of Scientific Thought in the German Genetics Community,* University of Chicago Press, Chicago, 1993. Among the many original publications on this subject, in particular see Sol Spiegelman and M. D. Kamen, "Genes and Nucleoproteins in the Synthesis of Enzymes," *Science,* vol. 104, 1946, pp. 581–584. The question of whether the nucleus or the cytoplasm determines inheritance and guides embryonic development had been posed at the end of the nineteenth century. The supporters of the cytoplasm also generally believed that development could be influenced by the external medium. The form of the cytoplasm does in fact depend on the surrounding medium: Garland E. Allen, "Morgan's Background and the Revolt from Descriptive and Speculative Biology," in T. J. Horder, J. A. Witkowski, and C. C. Wylie (eds.), *A History of Embryology,* Cambridge University Press, Cambridge, 1986, pp. 116–146.
22. Francis H. C. Crick, "The Present Position of the Coding Problem," *Brookhaven Symposia,* vol. 12, 1959, pp. 35–39; cited in Judson, *The Eighth Day of Creation,* p. 346.
23. See, for example, Gerard Hurwitz, Ann Bresler, and Renata Diringer, "The

Enzymic Incorporation of Ribonucleotides into Polyribonucleotides and the Effect of DNA," *Biochem. Biophys. Res. Comm.*, vol. 3, 1960, pp. 15–19.

24. James D. Watson, *The Double Helix: A Personal Account of the Discovery of the Structure of DNA*, Weidenfeld and Nicolson, London, 1968; rev. ed. by Gunther S. Stent, Weidenfeld and Nicolson, London, 1981, pp. 67–69.
25. Alexander Rich, "An Analysis of the Relation between DNA and RNA," *Ann. N.Y. Acad. Sci.*, vol. 81, 1959, pp. 709–72; and "A Hybrid Helix Containing Both Deoxyribose and Ribose Polynucleotides and Its Relation to the Transfer of Information between the Nucleic Acids," *Proc. Natl. Acad. Sci. USA*, vol. 46, 1960, pp. 1044–1053.
26. Elliot Volkin, "The Function of RNA in T2 Infected Bacteria," *Proc. Natl. Acad. Sci. USA*, vol. 46, 1960, pp. 1336–1349.
27. Mahlon B. Hoagland, "Nucleic Acids and Proteins," *Scientific American*, vol. 201, December 1959, pp. 55–61.
28. Masayasu Nomura, Benjamin D. Hall, and Sol Spiegelman, "Characterization of RNA, Synthesized in *Escherichia coli* after Bacteriophage T2 Infection," *J. Mol. Biol.*, vol. 2, 1960, pp. 306–326; Dario Giacomoni, "The Origin of DNA: RNA Hybridization," *J. Hist. Biol.*, vol. 26, 1993, pp. 89–107; Benjamin D. Hall and S. Spiegelman, "Sequence Complementarity of T2-DNA and T2-specific RNA," *Proc. Natl. Acad. Sci. USA*, vol. 47, 1961, pp. 137–146.
29. J. Marmur and D. Lane, "Strand Separation and Specific Recombination in Deoxyribonucleic Acids: Biological Studies," *Proc. Natl. Acad. Sci. USA*, vol. 46, 1960, pp. 453–461; Paul Doty, J. Marmur, J. Eigner, and C. Schildkraut, "Strand Separation and Specific Recombination in Deoxyribonucleic Acids: Physical Chemical Studies," *Proc. Natl. Acad. Sci. USA*, vol. 46, 1960, pp. 461–476.

14. The French School

1. Gunther S. Stent, "That Was the Molecular Biology That Was," *Science*, vol. 160, 1968, pp. 390–395.
2. Mirko D. Grmek and Bernardino Fantini, "Le rôle du hasard dans la naissance du modèle de l'opéron," *Revue d'histoire des sciences*, vol. 35, 1982, pp. 193–215.
3. Horace F. Judson, *The Eighth Day of Creation: The Makers of the Revolution in Biology*, Simon and Schuster, New York, 1979, chap. 7; Bernardino Fantini (ed.), *Jacques Monod: pour une éthique de la connaissance*, La Découverte, Paris, 1988; Patrice Debré, *Jacques Monod*, Flammarion, Paris, 1996.
4. Jacques Monod, "From Enzymatic Adaptation to Allosteric Transitions," *Science*, vol. 154, 1966, pp. 475–483; Jean-Paul Gaudillière, "J. Monod, S. Spiegel-

man et l'adaptation enzymatique. Programmes de recherche, cultures locales et traditions disciplinaires," *Hist. Phil. Life Sci.,* vol. 14, 1992, pp. 23–71.

5. As already suggested before the war by Marjory Stephenson. See Robert E. Kohler, "Innovation in Normal Science: Bacterial Physiology," *Isis,* vol. 76, 1985, pp. 162–181.
6. Jacques Monod, "The Phenomenon of Enzymatic Adaptation and Its Bearings on Problems of Genetics and Cellular Differentiation," *Growth Symposium,* vol. 11, 1947, pp. 223–289.
7. Benno Müller-Hill, *The "lac" Operon: A Short History of a Genetic Paradigm,* Walter De Gruyter, Berlin and New York, 1996.
8. Michel Morange, "L'œuvre scientifique de Jacques Monod," *Fundamenta Scientiae,* vol. 3, 1982, pp. 396–404.
9. Alvin M. Pappenheimer, Jr., "Qu'est donc devenue Pz?" in André Lwoff and Agnès Ullmann, *Les Origines de la biologie moléculaire. Un hommage à Jacques Monod,* "Études Vivantes," Paris, 1980, pp. 55–60.
10. Monod, "From Enzymatic Adaptation to Allosteric Transitions."
11. Melvin Cohn, Jacques Monod, Martin R. Pollock, Sol Spiegelman, and Roger Y. Stanier, "Terminology of Enzyme Formation," *Nature,* vol. 172, 1953, pp. 1096–1097.
12. Monod, who had been a communist during and after the war, broke with the Party after the Lysenko affair. In an article published on September 15, 1948, in the newspaper *Combat,* he denied that Lysenko and Mitchurin's theories had any scientific value whatsoever. For five years the attitude of the Communist Party did not change at all. The Mendelian theory of heredity and the respective merits of neo-Darwinism and neo-Lamarckism continued to fuel discussions among French left intellectuals. The question was not settled, and Monod felt the need to reaffirm publicly his rejection of all neo-Lamarckian ideas. See Joël and Dan Kottek, *L'Affaire Lyssenko,* Éditions Complexe, Brussels, 1986; Dominique Lecourt, *Proletarian Science?* New Left Books, London, 1976; Z. A. Medvedev, *The Rise and Fall of T. D. Lysenko,* Columbia University Press, New York, 1969.
13. Thomas D. Brock, *The Emergence of Bacterial Genetics,* Cold Spring Harbor Laboratory Press, Cold Spring Harbor, 1990.
14. Charles Galperin, "Le bactériophage, la lysogénie et son déterminisme génétique," *Hist. Phil. Life Sci.,* vol. 9, 1987, pp. 175–224; Charles Galperin, "Génétique et microbiologie: les problèmes de la lysogénie, 1925–1950," in Jean-Louis Fischer and William H. Schneider, *Histoire de la génétique, pratique, techniques et théories,* ARPEM et Sciences en situation, Paris, 1990, pp. 209–230.
15. François Jacob, *The Statue Within: An Autobiography,* Basic Books, New York, 1988.
16. Charles Galperin, "La lysogénie et les promesses de la génétique bactérienne,"

in *L'Institut Pasteur: contributions à son histoire,* La Découverte, Paris, 1991, pp. 198–206.

17. Grmek and Fantini, "Le rôle du hasard."
18. Arthur B. Pardee, François Jacob, and Jacques Monod, "The Genetic Control and Cytoplasmic Expression of 'Inducibility' in the Synthesis of β-galactosidase by *E. coli,*" *J. Mol. Biol.,* vol. 1, 1959, pp. 165–178.
19. François Jacob and Jacques Monod, "Genetic Regulatory Mechanisms in the Synthesis of Proteins," *J. Mol. Biol.,* vol. 3, 1961, pp. 318–356. For a description of this research, see Judson, *The Eighth Day of Creation,* chap. 7. See also Jean-Paul Gaudillière, "Biologie Moléculaire et biologistes dans les années soixante: la naissance d'une discipline. Le cas français," doctoral diss., Université de Paris-VII, Paris, 1991; Jean-Paul Gaudillière, "J. Monod, S. Spiegelman et l'adaptation enzymatique: programmes de recherche, cultures locales et traditions disciplinaires," *Hist. Phil. Life Sci.,* vol. 14, 1992, pp. 23–27; Grmek and Fantini, "Le rôle du hasard"; Kenneth Schaffner, "Logic of Discovery and Justification in Regulatory Genetics," *Stud. Hist. Phil. Sci.,* vol. 4, 1974, pp. 349–385.
20. Jacques Monod, "An Outline of Enzyme Induction," *Recueil des travaux chimiques des Pays-Bas,* vol. 77, 1958, p. 569.
21. François Jacob, "Genetic Control of Viral Function," *Harvey Lectures,* 1958–1959, Academic Press, New York, 1960, pp. 1–39.
22. Monod, "From Enzymatic Adaptation to Allosteric Transitions"; Jacob, *The Statue Within.*
23. H. Edwin Umbarger, "Evidence for a Negative-Feedback Mechanism in the Biosynthesis of Isoleucine," *Science,* vol. 123, 1956, p. 848.
24. Bernard T. Feld and Gertrud Weiss Szilard, *The Collected Papers of Leo Szilard: Scientific Papers,* MIT Press, Cambridge, Mass., 1972; William Lanouette and Bela Silard, *Genius in the Shadows: A Biography of Leo Szilard, the Man Behind the Bomb,* Charles Scribner's Sons, New York, 1992.
25. Michel Morange, "Le concept de gène régulateur," in Fischer and Schneider, *Histoire de la génétique,* pp. 271–291.
26. François Jacob and Jacques Monod, "Gènes de structure et gènes de régulation dans la biosynthèse des protéines," *C. R. Acad. Sci.* Paris, vol. 249, 1959, pp. 1282–1284.
27. François Jacob and Jacques Monod, "On the Regulation of Gene Activity," *Cold Spring Harbor Symp. Quant. Biol.,* vol. 26, 1961, pp. 193–211.
28. Jacques Monod, "Remarques conclusives du colloque: Basic Problems in Neoplastic Disease," in Bernardino Fantini (ed.), *Jacques Monod: pour une éthique de la connaissance,* La Découverte, Paris, 1988, pp. 79–96.
29. Evelyn Fox Keller, *A Feeling for the Organism: The Life and Work of Barbara McClintock,* W. H. Freeman and Co., New York, 1983; Nina Fedoroff and David Botstein (eds.), *The Dynamic Genome: Barbara McClintock's Ideas in*

the Century of Genetics, Cold Spring Harbor Laboratory Press, Cold Spring Harbor, 1991.

30. François Jacob, "Comments," *Cancer Research,* vol. 20, 1960, pp. 695–697.
31. Elie L. Wollman and François Jacob, *La sexualité des bactéries,* Masson, Paris, 1959 (translated as *Sexuality and the Genetics of Bacteria,* Academic Press, New York, 1961). The change undoubtedly took place in 1961. In the English version of the book, Jacob and Wollman were already much more cautious, as shown by the change in the title of the corresponding section: "Les épisomes et la différenciation cellulaire" ("Episomes and Cellular Differentiation") became "Episomes and Cellular Regulation."
32. Jacques Monod and François Jacob, "General Conclusions: Teleonomic Mechanisms in Cellular Metabolism, Growth and Differentiation," *Cold Spring Harbor Symp. Quant. Biol.,* vol. 26, 1961, pp. 389–401.
33. François Jacob and Jacques Monod, "Genetic Repression, Allosteric Inhibition and Cellular Differentiation," in *Cytodifferential and Macromolecular Synthesis,* Academic Press, New York, 1963, pp. 30–64.
34. Edward Yoxen, "Where Does Schrödinger's 'What Is Life?' Belong in the History of Molecular Biology?" *Hist. Sci.,* vol. 17, 1979, pp. 17–52.
35. Michel Morange, "Le concept de gène régulateur," in Fischer and Schneider, *Histoire de la génétique.*
36. Jacob does not recall there being any such direct influence (François Jacob, personal communication).
37. Claude Debru, *L'Esprit des protéines: histoire et philosophie biochimiques,* Hermann, Paris, 1983, chaps. 5 and 6.
38. Angela N. H. Creager and Jean-Paul Gaudillière, "Meanings in Search of Experiments or *Vice-Versa:* The Invention of *Allosteric Regulation* in Paris and Berkeley (1959–1969)," *Hist. Stud. Phys. Biol. Sci.,* vol. 27, 1996, pp. 1–89.
39. Henri Buc, "Dame Nature façonnant les enzymes de régulation," in Lwoff and Ullman, *Les Origines de la biologie moléculaire;* Daniel E. Koshland, Jr., "Quelques souvenirs de Jacques Monod," in Lwoff and Ullman, *Les Origines de la biologie moléculaire.* The scale and intensity of these controversies might seem surprising. They partly continued the old debates between Darwinians and neo-Lamarckians. Koshland's model was given a neo-Lamarckian interpretation by Carl C. Lindegren, who played a key role in the genetic analysis of *Neurospora* (see Chapter 2) and yeast, but who also became a fervent opponent of "Morganian" genetics. See Carl C. Lindegren, "Lamarckian Proteins," *Nature,* vol. 198, 1963, p. 1224.
40. Jean-Pierre Changeux subsequently turned to the study of the acetylcholine receptor, which he interpreted according to allosteric theory. He thus paved the way for molecular biologists to enter neurobiology: Jean-Pierre Changeux, "Allosteric Proteins: From Regulatory Enzymes to Receptors—Personal Recollections," *Bioessays,* vol. 15, 1993, 625–634; Thierry Heidmann

and Jean-Pierre Changeux, "Structural and Functional Properties of the Acetylcholine Receptor Protein in Its Purified and Membrane-Bound States," *Ann. Rev. Biochem.*, vol. 47, 1978, pp. 317–357.

41. Jacques Monod, Jeffries Wyman, and Jean-Pierre Changeux, "On the Nature of Allosteric Transitions: A Plausible Model," *J. Mol. Biol.*, vol. 12, 1965, pp. 88–118.
42. Jacques Monod wrote the preface to the first French edition of Popper's *The Logic of Scientific Discovery* (Hutchinson, London, 1959), translated as *La Logique de la découverte scientifique*, Payot, Paris, 1973.
43. François Jacob, Sydney Brenner, and François Cuzin, "On the Regulation of DNA Replication in Bacteria," *Cold Spring Harbor Symp. Quant. Biol.*, vol. 28, 1963, pp. 329–348.
44. François Jacob, "Biologie moléculaire, la prochaine étape," *Atomes*, no. 271, 1969, pp. 748–750; François Jacob, *La souris, la mouche et l'homme*, Odile Jacob, Paris, 1997, chap. 3. Other scientists in his laboratory adopted a more cautious strategy and began to study cellular differentiation in an amoeba, *Dictyostelium discoideum.*
45. Jean-Paul Gaudillière, "Catalyse enzymatique et oxydations cellulaires: l'œuvre de Gabriel Bertrand et son héritage," in *L'Institut Pasteur: contributions à son histoire*, pp. 118–136.
46. Yvette Conry, *L'Introduction du darwinisme en France au XIXe siècle*, Vrin, Paris, 1974.
47. Denis Buican, *Histoire de la génétique et de l'évolutionnisme en France*, PUF, Paris, 1984; Jan Sapp, *Beyond the Gene: Cytoplasmic Inheritance and the Struggle for Authority in Genetics*, Oxford University Press, New York, 1987, chap. 5.
48. The remarkable work by Philippe L'Héritier and Georges Teissier on experimental populations of *Drosophila* in the 1930s must, however, be mentioned (Jean Gayon, *Darwin et l'après-Darwin: une histoire de l'hypothèse de sélection naturelle*, Kimé, Paris, 1992, pp. 379–384; English translation: *Darwinism's Struggle for Survival: Heredity and the Hypothesis of Natural Selection*, Cambridge University Press, Cambridge, England, forthcoming).
49. Richard M. Burian, Jean Gayon, and Doris Zallen, "The Singular Fate of Genetics in the History of French Biology, 1900–1940," *J. Hist. Biol.*, vol. 21, 1988, pp. 357–402.
50. Richard M. Burian and Jean Gayon, "Genetics after World War II: The Laboratories at Gif," *Cahiers pour l'histoire du CNRS*, vol. 7, 1990, pp. 25–48.
51. Richard M. Burian and Jean Gayon, "Un évolutionniste bernardien à l'Institut Pasteur? Morphologie des ciliés et évolution physiologique dans l'œuvre d'André Lwoff," in *L'Institut Pasteur: contributions à son histoire*, pp. 165–186.
52. Ibid.
53. Sapp, *Beyond the Gene*, chap. 5.

15. Normal Science

1. Thomas S. Kuhn, *The Structure of Scientific Revolutions,* University of Chicago Press, Chicago, 1970.
2. Gunther S. Stent, *The Coming of the Golden Age: A View of the End of Progress,* Natural History Press, Garden City, N.Y., 1969.
3. Ernest P. Fischer and Carol Lipson, *Thinking about Science: Max Delbrück and the Origins of Molecular Biology,* W. W. Norton, New York, 1988.
4. Yoshiki Hotta and Seymour Benzer, "Mapping of Behaviour in *Drosophila* Mosaics," *Nature,* vol. 240, 1972, pp. 527–535.
5. Stent, *The Coming of the Golden Age;* Jacob, *La logique du vivant.*
6. Jacques Monod, *Chance and Necessity,* Collins, London, 1972, p. 12.
7. See, for example, Pierre-Henri Simon, *Questions aux savants,* Le Seuil, Paris, 1969; Marc Beigbeder, *Le Contre-Monod,* Grasset, Paris, 1972; Madeleine Barthélémy-Madaule, *L'Idéologie du hasard et de la nécessité,* Le Seuil, Paris, 1972.
8. The intensity of the debate that accompanied the publication of *Le Hasard et la Nécessité* is explained more by Monod's personality and his place in French society than by the book's content. Monod was the only scientist among the intellectuals who played an important part in French life between the Second World War and the end of the 1970s. Very active in the Communist resistance during the war, but having spectacularly broken with the Communist Party over the Lysenko affair, Monod remained a "leftist," close to Albert Camus, whose work he admired profoundly. Monod was one of the scientists close to the circle of Pierre Mendès France, who tried to renovate the French university and research systems in the late 1950s, without much success. Basking in the official recognition that followed the Nobel Prize, and as the director of the Pasteur Institute from 1971 on, Monod often intervened in public affairs. He participated in the French Family Planning Movement in the campaign for birth control and the liberalization of contraception, and was also active in protests demanding the legalization of abortion, the right to euthanasia, and support for imprisoned Soviet scientists. His positions were always clear, but his opponents sometimes thought they were marked by contempt. This explains why many of the debates he took part in were particularly bitter. See *Les intellectuels français,* Jacques Julliard and Michel Winock (eds.), Le Seuil, Paris, 1996, pp. 800–801; Patrice Debré, *Jacques Monod,* Flammarion, Paris, 1996; Service des Archives de l'Institut Pasteur, Fonds Monod. Monod was also a musician. He was torn for a long time between scientific research and a career in music. He was also interested in literature and even wrote a play, *Le puits de Syène* (1964).

9. Maxime Schwartz, "Une autre voie," in André Lwoff and Agnès Ullman, *Les Origines de la biologie moléculaire. Un hommage à Jacques Monod,* "Études Vivantes," Paris, 1980, pp. 177–184.
10. David Baltimore, "Viral RNA-Dependent DNA Polymerase," *Nature,* vol. 226, 1970, pp. 1209–1211; Howard M. Temin and S. Mizutani, "RNA-Dependent DNA Polymerase in Virions of Rous Sarcoma Virus," *Nature,* vol. 226, 1970, pp. 1211–1213; Lindley Darden, "Exemplars, Abstractions, and Anomalies: Representations and Theory Change in Mendelian and Molecular Genetics," in James G. Lennox and Gereon Walters (eds.), *Philosophy of Biology,* University of Konstanz Press, Konstanz, Germany, and University of Pittsburgh Press, Pittsburgh, Penn., 1995, pp. 137–158.
11. Francis H. Crick, "On Protein Synthesis," *Symp. Soc. Exptl. Biol.,* vol. 12, pp. 138–163.
12. Francis Crick, *What Mad Pursuit: A Personal View of Scientific Discovery,* Basic Books, New York, 1988.
13. Renato Dulbecco, "Production of Plaques in Monolayer Tissue Cultures by Single Particles of an Animal Virus," *Proc. Natl. Acad. Sci. USA,* vol. 38, 1952, pp. 747–752. For a review, see Renato Dulbecco, "From the Molecular Biology of Oncogenic DNA Viruses to Cancer," *Science,* vol. 192, 1976, pp. 437–440; R. Dulbecco, *Scienza, Vita e Aventura,* Sperling and Kupfer, Milan, 1989; Daniel J. Kevles, "Renato Dulbecco and the New Animal Virology: Medicine, Methods and Molecules," *J. Hist. Biol.,* vol. 26, 1993, pp. 409–442.
14. "Central Dogma Reversed," *Nature,* vol. 226, 1970, pp. 1198–1199.
15. Howard M. Temin, "The Protovirus Hypothesis," *J. Natl. Cancer. Inst.,* vol. 46, 1971, pp. iii–viii. Temin thus proposed a "Lamarckian" view of cell functioning. The genetic content of a cell reflected its "state" of functioning. Since cells could modify the expression of their genes as a function of external signals—as Jacob and Monod had shown—the genetic make-up of a cell depended on this external medium.
16. S. J. Singer and Garth L. Nicholson, "The Fluid Mosaic Model of the Structure of Cell Membranes," *Science,* vol. 175, 1972, pp. 720–721.
17. The first journal of molecular biology (appropriately named the *Journal of Molecular Biology*) was started in 1959 by John Kendrew. A recent selection of papers by Brenner gives an idea of the importance of this journal in the publication of results in molecular biology. See Sydney Brenner, *Molecular Biology: A Selection of Papers,* Academic Press, London, 1989. The initiative for this journal, however, did not come from molecular biologists: Robert C. Olby, "The Molecular Revolution in Biology," in R. C. Olby, G. N. Cantor, J. R. R. Christie, and M. J. S. Hodge (eds.), *Companion to the History of Modern Science,* Routledge, London, 1990, p. 507. The rapid expansion of molecular biology undoubtedly explains why it did not have the time to crystallize into a genuinely new discipline, but fused with pre-existing disciplines.

18. Jean-Paul Gaudillière, "Biologie moléculaire et biologistes dans les années soixante: la naissance d'une discipline. Le cas français," doctoral diss., Université de Paris-VII, Paris, 1991; Jean-Paul Gaudillière, "Molecular Biology in the French Tradition? Redefining Local Traditions and Disciplinary Patterns," *J. Hist. Biol.,* vol. 26, 1993, pp. 473–498; Jean-Paul Gaudillière, "Chimie biologique ou chimie moléculaire? La biochimie au CNRS dans les années soixante," *Cahiers pour l'histoire du CNRS,* vol. 7, 1990, pp. 91–147; Jean-Paul Gaudillière, "Molecular Biologists, Biochemists and Messenger RNA: The Birth of a Scientific Network," *J. Hist. Biol.,* vol. 29, 1996, pp. 417–445; Xavier Polanco, "La mise en place d'un réseau scientifique, les rôles du CNRS et de la DGRST dans l'institutionnalisation de la biologie moléculaire en France (1960–1970)," *Cahiers pour l'histoire du CNRS,* vol. 7, 1990, pp. 49–90. For an international view of this "seizure of power" by the molecular biologists and their confrontation with the biochemists, see Pnina G. Abir-Am, "The Politics of Macromolecules: Molecular Biologists, Biochemists, and Rhetoric," *Osiris,* vol. 7, 1992, pp. 164–191.
19. Bernd Gutte and R. B. Merrifield, "The Synthesis of Ribonuclease A," *J. Biol. Chem.,* vol. 246, 1971, pp. 1922–1941.
20. The first protein structures (myoglobin and hemoglobin, respectively) had been determined earlier: John C. Kendrew, R. E. Dickerson, B. E. Strandberg, R. G. Hart, D. R. Davies, D. C. Philipps, and V. C. Shore, "A Three-Dimensional Model of the Myoglobin Molecule Obtained by X-Ray Analysis," *Nature,* vol. 181, 1958, pp. 662–666; Max F. Perutz, M.G. Rossmann, Ann F. Cullis, Hillary Muirhead, George Will, and A.C.T. North, "Structure of Hemoglobin: A Three-Dimensional Fourier Synthesis at 5.5Å Resolution Obtained by X-Ray Analysis," *Nature,* vol. 185, 1960, pp. 416–422. Over the next few years, both an improved definition of these structures and the determination of new three-dimensional protein structures (in particular of enzymes) became possible (see below). The determination of these structures was rendered possible by Perutz's 1951 solution to the phase problem (see Chapter 11), by continuous advances in the quantitative measurement of diffraction patterns, and by computer processing of the experimental data.
21. The first enzyme whose catalytic mechanism was determined using crystallography was lysozyme (1967): C. C. F. Blake, L. N. Johnson, G. A. Mair, A. C. T. North, D. C. Philipps, and V. R. Sarma, "Crystallographic Studies of the Activity of Hen Egg-White Lysozyme," *Proc. Roy. Soc. B,* London, vol. 167, 1967, pp. 378–388. The structure and mechanism of action of chymotrypsin and carboxypeptidase were determined shortly afterwards.
22. Okazaki also showed that DNA replication was a discontinuous process: Reiji Okazaki, Tuneko Okazaki, Kiwako Sakabe, Kazunori Sugimoto, and Akio Sugino, "Mechanism of DNA Chain Growth I. Possible Discontinuity and Unusual Secondary Structure of Newly Synthesized Chains," *Proc. Natl. Acad.*

Sci. USA, vol. 59, 1968, pp. 598–605. DNA replication turned out to be an extremely complicated process, requiring the involvement of many protein factors: Arthur Kornberg, *For the Love of Enzymes: The Odyssey of a Biochemist,* Harvard University Press, Cambridge, Mass., 1989.

23. Richard R. Burgess, Andrew A. Travers, John J. Dunn, and Ekkehard K. F. Bautz, "Factors Stimulating Transcription by RNA Polymerase," *Nature,* vol. 221, 1969, pp. 43–46.
24. Sung Hou Kim, Gary Quigley, F. L. Suddath, and Alexander Rich, "High Resolution X-Ray Diffraction Patterns of Crystalline Transfer RNA That Show Helical Regions," *Proc. Natl. Acad. Sci. USA,* vol. 68, 1971, pp. 841–845.
25. Robert W. Holley, Jean Apgar, George A. Everett, James T. Madison, Mark Marquisee, Susan H. Merrill, John Robert Penswick, and Ada Zamir, "Structure of a Ribonucleic Acid," *Science,* vol. 147, 1965, pp. 1462–1465.
26. Jim Shapiro, Lorne Machattie, Larry Eron, Garrett Ihler, Karin Ippen, and Jon Beckwith, "Isolation of Pure *Lac* Operon DNA," *Nature,* vol. 224, 1969, pp. 768–774. For the isolation of the first genes of higher organisms (ribosomal genes), see the review by Max L. Birnstiel, "Gene Isolation is 25 Years Old This Month," *TIG,* vol. 6, 1990, pp. 380–381.
27. K. L. Agarwal, H. Buchi, M. H. Caruthers, N. Gupta, H. G. Khorana, K. Kleppe, A. Kumar, E. Ohtsuka, V. L. Rajbhandary, J. H. Van de Sande, V. Sgaramella, H. Weber, and T. Yamada, "Total Synthesis of the Gene for an Alanine Transfer Ribonucleic Acid from Yeast," *Nature,* vol. 227, 1970, pp. 27–34.
28. François Jacob, "Biologie moléculaire, la prochaine étape," *Atomes,* no. 271, 1969, pp. 748–750.
29. Walter Gilbert and Benno Müller-Hill, "Isolation of the Lac Repressor," *Proc. Natl. Acad. Sci. USA,* vol. 56, 1966, pp. 1891–1898; Mark Ptashne, "Isolation of the Lambda Phage Repressor," *Proc. Natl. Acad. Sci. USA,* vol. 57, 1967, pp. 306–313.
30. Gaudillière, *Biologie Moléculaire et biologistes dans les années soixante,* pp. 211–212.
31. The first observations of the instability of nuclear RNA had been made in 1963 by Henry Harris: H. Harris, H. W. Fisher, A. Rodgers, T. Spencer, and J. W. Watts, "An Examination of the Ribonucleic Acids in the HeLa Cells with Special Reference to Current Theory about the Transfer of Information from Nucleus to Cytoplasm," *Proc. Roy. Soc. London, Ser. B,* vol. 157, 1963, pp. 177–198.
32. Roy J. Britten and Eric H. Davidson, "Gene Regulation for Higher Cells: A Theory," *Science,* vol. 165, 1969, pp. 349–357.
33. The crisis felt by molecular biologists has been well described by François Gros: "The history of the development of ideas—in science as in art—shows that it is dangerous to push the exploitation of a concept or of a methodology to its limits, because a feeling of saturation will tend to replace the satisfaction that accompanied the initial phase. Not only did researchers—including the finest minds in the discipline—start to wonder about the future of molecular

biology, but by the beginning of the 1970s the whole discipline went into crisis. Of course, impelled by the "dynamic" of its successes, it carried on and continued to obtain some important results, but originality was not always a hallmark of the science. One has to admit that research was treading water and the heart had gone out of it. This "low point" was accompanied, if not by a genuine anxiety, at least by a questioning that was not totally unlike anxiety." François Gros, *Les secrets du gène,* Odile Jacob, Paris, 1986, p. 167.

34. See, for example, Jacqueline Djian (ed.), *La médecine moléculaire,* Robert Laffont, Paris, 1970.
35. Christiane Sinding, *Le clinicien et le chercheur: des grandes maladies de carence à la médecine moléculaire* (1880–1980), Presses Universitaires de France, Paris, 1991, chap. 7.
36. Linus Pauling, Harvey A. Itano, S. J. Singer, and Ibert C. Wells, "Sickle Cell Anemia, a Molecular Disease," *Science,* vol. 110, 1949, pp. 543–548.
37. Gaudillière, *Biologie moléculaire et biologistes dans les années soixante.*
38. This retreat can also be justified scientifically. Regulatory mechanisms controlling enzyme activity, such as phosphorylation by protein kinases, do not exist (or rarely exist) in bacteria. It was quite reasonable to imagine that the complexity of higher organisms was located as much (if not more) at the biochemical level as at the level of gene expression. For example, a hormone's binding to its receptor frequently causes the synthesis of a small molecule called cyclic AMP in the cell, which in turn activates a protein kinase. This pathway of intracellular signaling was particularly well studied. See Earl W. Sutherland, "Studies on the Mechanism of Hormone Action," *Science,* vol. 177, 1972, pp. 401–408.
39. M. C. Niu, "Thymus Ribonucleic Acid and Embryonic Differentiation," *Proc. Natl. Acad. Sci. USA,* vol. 44, 1958, pp. 1264–1274; Jean-Paul Gaudillière, "Un code moléculaire pour la différenciation cellulaire: la controverse sur les transferts d'ARN informationnel (1955–1973) et les étapes de diffusion du paradigme de la biologie moléculaire," *Fundamenta Scientiae,* vol. 9, 1988, pp. 429–467.
40. Marvin Fishman, R. A. Hammerstrom, and V. P. Bond, "In Vitro Transfer of Macrophage RNA to Lymph Node Cells," *Nature,* vol. 198, 1963, pp. 549–551; Gaudillière, "Une code moléculaire pour la différenciation cellulaire."
41. Michel Morange, "La recherche d'un code moléculaire de la mémoire," *Fundamenta Scientiae,* vol. 6, 1985, pp. 65–80.
42. The controversy over the possibility of learning in planarian worms was studied by G. D. L. Travis, "Replicating Replication? Aspects of the Social Construction of Learning in Planarian Worms," *Soc. Stud. Sci.,* vol. 11, 1981, pp. 11–32.
43. W. L. Byrne, D. Samuel, E. L. Bennett, M. R. Rosenzweig, E. Wasserman, A. R. Wagner, F. Gardner, R. Galambos, B. D. Berger, D. L. Margules, R. L. Fenichel, L. Stein, J. A. Corson, H. E. Enesco, S. L. Chourouer, C. E. Holt III,

P. H. Schiller, L. Chiappetta, M. E. Jarvik, R. C. Leaf, J. D. Dutcher, Z. P. Horovitz, and P. L. Carson, "Memory Transfer," *Science,* vol. 153, 1966, pp. 658–659.

44. D. F. Tate, L. Galvan, and George Ungar, "Isolation and Identification of Two Learning-Induced Peptides," *Pharmacol. Biochem. Behav.,* vol. 5, 1976, pp. 441–448.
45. In 1972, *Nature* published an article by Ungar and coworkers describing a peptide responsible for the fear reaction to dark—"scotophobin." See George Ungar, D. M. Desiderio, and W. Parr, "Isolation, Identification and Synthesis of a Specific Behaviour-Inducing Brain Peptide," *Nature,* vol. 238, 1972, pp. 198–210. In the same issue the journal published a critical review written by the article's referees and a response by the authors.
46. Michel Morange, "Science et effet de mode," *L'État des sciences et des techniques,* under the direction of Nicolas Witkowski, La Découverte, Paris, 1991, pp. 453–454.

16. Genetic Engineering

1. Oswald T. Avery, Colin M. MacLeod, and Maclyn McCarty, "Studies on the Chemical Nature of the Substance Inducing Transformation of Pneumococcal Types," *J. Exp. Med.,* vol. 79, 1944, pp. 137–158.
2. Edward L. Tatum, "A Case History in Biological Research," *Science,* vol. 129, 1959, pp. 1711–1715.
3. Joshua Lederberg, "Genetics," *Encyclopaedia Britannica, Yearbook of Science and the Future,* 1969, p. 321.
4. Elizabeth H. Szybalska and Waclaw Szybalski, "Genetics of Human Cell Lines, IV. DNA-Mediated Heritable Transformation of a Biochemical Trait," *Proc. Natl. Acad. Sci. USA,* vol. 48, 1962, pp. 2026–2034.
5. Susan Wright, "Recombinant DNA Technology and Its Social Transformation, 1972–1982," *Osiris,* vol. 2, 1986, pp. 303–360.
6. For an overview of this work, see Werner Arber, "Promotion and Limitation of Genetic Exchange," *Science,* vol. 205, 1979, pp. 361–365.
7. Hamilton O. Smith and K. W. Wilcox, "A Restriction Enzyme from Hemophilus Influenzae I. Purification and General Properties," *J. Mol. Biol.,* vol. 51, 1970, pp. 379–391; Thomas J. Kelly and Hamilton O. Smith, "A Restriction Enzyme from Hemophilus Influenzae II. Base Sequence of the Recognition Site," *J. Mol. Biol.,* vol. 51, 1970, pp. 393–409. The first restriction enzyme was purified by Meselson and Yuan: Matthew Meselson and Robert Yuan, "DNA Restriction Enzyme from *E. coli,*" *Nature,* vol. 217, 1968, pp. 1110–1114. This enzyme had a low cleavage specificity, which meant that it was not particularly useful.

8. Kathleen Danna and Daniel Nathans, "Specific Cleavage of Simian Virus 40 DNA by Restriction Endonuclease of Hemophilus Influenzae," *Proc. Natl. Acad. Sci. USA,* vol. 68, 1971, pp. 2913–2917.
9. David A. Jackson, Robert H. Symons, and Paul Berg, "Biochemical Method for Inserting New Genetic Information into DNA of Simian Virus 40: Circular SV40 Molecules Containing Lambda Phage Genes and the Galactose Operon of *Escherichia coli,*" *Proc. Natl. Acad. Sci. USA,* vol. 69, 1972, pp. 2904–2909.
10. Joshua Lederberg, "Genetics of Bacteria," Grant Application to the National Institutes of Health, no. A1 05160-11, December 20, 1967. Cited in Wright, "Recombinant DNA Technology and Its Social Transformation, 1972–1982," p. 310.
11. Peter Lobban and A. D. Kaiser, "Enzymatic End to End Joining of DNA Molecules," *J. Mol. Biol.,* vol. 78, 1973, pp. 453–471. Lobban's results were published after a year's delay, which tended to hide the fact that the two experiments were in fact carried out simultaneously. Lobban's thesis advisor, A. D. Kaiser, preferred that Lobban correct his thesis rather than publish his results. Reported by Arthur Kornberg, *For the Love of Enzymes: The Odyssey of a Biochemist,* Harvard University Press, Cambridge, Mass., 1989, p. 275 et seq.
12. Gunther S. Stent, "Prematurity and Uniqueness in Scientific Discovery," *Scientific American,* vol. 227, December 1972, pp. 84–93.
13. A chronology of the principal discoveries that gave rise to genetic engineering can be found in James D. Watson and John Tooze, *A Documentary History of Gene Cloning,* W. H. Freeman and Co., San Francisco, 1981; Jan Witkowski, "Fifty Years of Molecular Biology's Hall of Fame," *Life Science Job Trends,* vol. 2, no. 17, 1988, pp. 1–13. For the development and early applications of genetic engineering, see Stephen S. Hall, *Invisible Frontiers: The Race to Synthesize a Human Gene,* Atlantic Monthly Press, New York, 1987.
14. Janet E. Mertz and Ronald W. Davis, "Cleavage of DNA by RI Restriction Endonuclease Generates Cohesive Ends," *Proc. Natl. Acad. Sci. USA,* vol. 69, 1972, pp. 3370–3374.
15. Stanley N. Cohen, Annie C. Y. Chang, Herbert W. Boyer, and Robert B. Helling, "Construction of Biologically Functional Bacterial Plasmids *in vitro,*" *Proc. Natl. Acad. Sci. USA,* vol. 70, 1973, pp. 3240–3244. To make the plasmids enter the bacteria, these scientists used a calcium-chloride–based method developed earlier by Mohrt Mandel (M. Mandel and A. Higa, "Calcium Dependent Bacteriophage DNA Infection," *J. Mol. Biol.,* vol. 53, 1970, pp. 159–162).
16. Annie C. Y. Chang and Stanley N. Cohen, "Genome Construction between Bacterial Species in vitro: Replication and Expression of *Staphylococcus* Plasmid Genes in *Escherichia coli,*" *Proc. Natl. Acad. Sci. USA,* vol. 71, 1974, pp. 1030–1034.
17. John F. Morrow, Stanley N. Cohen, Annie C. Y. Chang, Herbert W. Boyer, Howard M. Goodman, and Robert B. Helling, "Replication and Transcription of Eukaryotic DNA in *Escherichia coli,*" *Proc. Natl. Acad. Sci. USA,* vol. 71, 1974, pp. 1743–1747.

18. Maxine Singer and Dieter Soll, "Guidelines for DNA Hybrid Molecules," *Science,* vol. 181, 1973, p. 1114.
19. Paul Berg et al., "Potential Biohazards of Recombinant DNA Molecules," *Proc. Natl. Acad. Sci. USA,* vol. 71, 1974, pp. 2593–2594; Paul Berg et al., "Potential Biohazards of Recombinant DNA Molecules," *Science,* vol. 185, 1974, p. 303; "NAS Ban on Plasmid Engineering," *Nature,* vol. 250, 1974, p. 175.
20. Paul Berg, David Baltimore, Sydney Brenner, Richard O. Roblin, and Maxine F. Singer, "Asilomar Conference on Recombinant DNA Molecules," *Science,* vol. 188, 1975, pp. 44–47.
21. Michael Ruse, "The Recombinant DNA Debate: A Tempest in a Test Tube?" in *Is Science Sexist?* D. Reidel Publishing Company, Dordrecht, Holland, 1981; Clifford Grobstein, *A Double Image of the Double Helix: The Recombinant DNA Debate,* W. H. Freeman and Co., San Francisco, 1979; Sheldon Krimsky, *Genetic Alchemy: The Social History of the Recombinant DNA Controversy,* MIT Press, Cambridge, Mass., 1982; Susan Wright, *Molecular Politics: Developing American and British Regulatory Policy for Genetic Engineering, 1972–1982,* University of Chicago Press, Chicago, 1994.
22. Hall, *Invisible Frontiers;* Marcel Blanc, *L'ère de la génétique,* La Découverte, Paris, 1986.
23. Paul Berg, "Dissections and Reconstructions of Genes and Chromosomes," *Science,* vol. 213, 1981, pp. 296–303.
24. Susan Wright, "Molecular Biology or Molecular Politics? The Production of Scientific Consensus on the Hazards of Recombinant DNA Technology," *Soc. Stud. of Sci.* vol. 16, 1986, pp. 593–620.
25. Argiris Efstratiadis, Fotis C. Kafatos, Allan M. Maxam, and Tom Maniatis, "Enzymatic In Vitro Synthesis of Globin Genes," *Cell,* vol. 7, 1976, pp. 279–288.
26. Tom Maniatis, Sim Gek Kee, Argiris Efstratiadis, and Fotis C. Kafatos, "Amplification and Characterization of a β-Globin Gene Synthesized In Vitro," *Cell,* vol. 8, 1976, pp. 163–182.
27. Pieter Wensink, David J. Finnegan, John E. Donelson, and David S. Hogness, "A System for Mapping DNA Sequences in the Chromosomes of *Drosophila melanogaster,*" *Cell,* vol. 3, 1974, pp. 315–325.
28. Tom Maniatis, Ross C. Hardison, Elizabeth Lacy, Joyce Lauer, Catherine O'Connel, Diana Quon, Gek Kee Sim, and Argiris Efstratiadis, "The Isolation of Structural Genes from Libraries of Eucaryotic DNA," *Cell,* vol. 15, 1978, pp. 687–701.
29. Michael Grunstein and David S. Hogness, "Colony Hybridization: A Method for the Isolation of Cloned DNAs That Contain a Specific Gene," *Proc. Natl. Acad. Sci. USA,* vol. 72, 1975, pp. 3961–3965.
30. In *Drosophila,* the existence of giant chromosomes in the salivary glands makes it easy to localize directly the position of a DNA fragment on the chromosomal map, by *in situ* hybridization. See Wensink et al. (note 27).

31. Hamilton O. Smith, "Nucleotide Sequence Specificity of Restriction Endonucleases," *Science,* vol. 205, 1979, pp. 455–462.
32. Phillip A. Sharp, Bill Sugden, and Joe Sambrook, "Detection of Two Restriction Endonuclease Activities in *Haemophilus parainfluenzae* Using Analytical Agarose–Ethidium Bromide Electrophoresis," *Biochemistry,* vol. 12, 1973, pp. 3055–3063. The ultracentrifugation method continued to be used in the preparation of plasmids: in the presence of ethidium bromide, plasmids have a density different from that of chromosomal DNA and can easily be separated on a cesium chloride gradient.
33. E. M. Southern, "Detection of Specific Sequences among DNA Fragments Separated by Gel Electrophoresis," *J. Mol. Biol.,* vol. 98, 1975, pp. 503–517; Dario Giacomoni, "The Origin of DNA:RNA Hybridization," *J. Hist. Biol.,* vol. 26, 1993, pp. 89–107.
34. James C. Alwine, David J. Kemp, and George R. Stark, "Method for Detection of Specific RNAs in Agarose Gels by Transfer to Diazobenzyloxymethyl-Paper and Hybridization with DNA Probes," *Proc. Natl. Acad. Sci. USA,* vol. 74, 1977, pp. 5350–5354.
35. Francisco Bolivar, Raymond L. Rodriguez, Mary C. Betlach, and Herbert W. Boyer, "Construction and Characterization of New Cloning Vehicles: I. Ampicillin-Resistant Derivatives of the Plasmid pMB9," *Gene,* vol. 2, 1977, pp. 75–93; Francisco Bolivar, Raymond L. Rodriguez, Patricia J. Greene, Mary C. Betlach, Herbert L. Heynecker, Herbert W. Boyer, Jorge H. Crosa, and Stanley Falkow, "Construction and Characterization of New Cloning Vehicles: II. A Multipurpose Cloning System," *Gene,* vol. 2, 1977, pp. 95–113.
36. John Collins and Barbara Hohn, "Cosmids: A Type of Plasmid Gene-Cloning Vector That Is Packageable *in vitro* in Bacteriophage λ Heads," *Proc. Natl. Acad. Sci. USA,* vol. 75, 1978, pp. 4242–4246.
37. Allan M. Maxam and Walter Gilbert, "A New Method for Sequencing DNA," *Proc. Natl. Acad. Sci. USA,* vol. 74, 1977, pp. 560–564. The Russian chemist Eugene Sverdlov had proposed a similar method four years earlier, but his work had not attracted any attention: E. D. Sverdlov, W. Monastyrskaya, A. V. Chestukhin, and E. I. Budowsky, "The Primary Structure of Oligonucleotides: Partial Apurination as a Method to Determine the Position of Purine and Pyrimidine Residues," *FEBS Letters,* vol. 33, 1973, pp. 15–17.
38. Frederick Sanger, S. Nicklen, and A. R. Coulson, "DNA Sequencing with Chain-Terminating Inhibitors," *Proc. Natl. Acad. Sci. USA,* vol. 74, 1977, pp. 5463–5467.
39. Clyde A. Hutchison III, Sandra Phillips, Marshall H. Edgell, Shirley Gillam, Patricia Jahnke, and Michael Smith, "Mutagenesis at a Specific Position in a DNA Sequence," *J. Biol. Chem.,* vol. 253, 1978, pp. 6551–6560.
40. M. J. Gait and R. C. Sheppard, "Rapid Synthesis of Oligodeoxyribonucleotides: A New Solid-Phase Method," *Nucleic Acid Research,* vol. 4, 1977, pp. 1135–1158.
41. The authors modified the technique that was already used in bacteria:

F. L. Graham and A. J. Van der Erb, "A New Technique for the Assay of Infectivity of Human Adenovirus DNA," *Virology*, vol. 52, 1973, pp. 456–467.

42. Michael Wigler, Raymond Sweet, Gek Kee Sim, Barbara Wold, Angel Pellicer, Elizabeth Lacy, Tom Maniatis, Saul Silverstein, and Richard Axel, "Transformation of Mammalian Cells with Genes from Prokaryotes and Eukaryotes," *Cell*, vol. 16, 1979, pp. 777–785.
43. Tom Maniatis, Ed. F. Fritsch, and Joe Sambrook, *Molecular Cloning: A Laboratory Manual*, Cold Spring Harbor Laboratory Press, Cold Spring Harbor, 1982. In laboratories, this book is often called "the recipe book," "the cookbook," or more respectfully, "the Bible." Joan H. Fujimura, "Constructing 'Do-Able' Problems in Cancer Research: Articulating Alignment," *Soc. Stud. Sci.*, vol. 17, 1987, pp. 257–293.
44. The generalization of data from one organism to the whole of biology was a strategy that was quite reasonable even if it often turned out to be fruitless. Lindley Darden, "Essay Review: Generalizations in Biology," *Studies Hist. Phil. Sci.*, vol. 27, 1996, pp. 409–419.
45. Maniatis et al., "Amplification and Characterization."
46. Axel Ullrich, John Shine, John Chirgwin, Raymond Pictet, Edmond Tischer, William J. Rutter, and Howard M. Goodman, "Rat Insulin Genes: Construction of Plasmids Containing the Coding Sequences," *Science*, vol. 196, 1977, pp. 1313–1319.
47. Peter H. Seeburg, John Shine, Joseph A. Martial, John D. Baxter, and Howard M. Goodman, "Nucleotide Sequence and Amplification in Bacteria of Structural Gene for Rat Growth Hormone," *Nature*, vol. 270, 1977, pp. 486–494.
48. John Shine, Peter H. Seeburg, Joseph A. Martial, John D. Baxter, and Howard M. Goodman, "Construction and Analysis of Recombinant DNA for Human Chorionic Somatomammotropin," *Nature*, vol. 270, 1977, pp. 494–499.
49. Tom Maniatis et al., "Amplification and Characterization."
50. Keiichi Itakura, Tadaaki Hirose, Roberto Crea, Arthur D. Riggs, Herbert Heynecker, Francisco Bolivar, and Herbert Boyer, "Expression in *Escherichia coli* of a Chemically Synthesized Gene for the Hormone Somatostatin," *Science*, vol. 198, 1977, pp. 1056–1063.
51. David V. Goeddel, Dennis G. Kleid, Francisco Bolivar, Herbert L Heynecker, Daniel G. Yansura, Roberto Crea, Tadaaki Hirose, Adam Kraszewski, Keiichi Itakura, and Arthur D. Riggs, "Expression in *Escherichia coli* of Chemically Synthesized Genes for Human Insulin," *Proc. Natl. Acad. Sci. USA*, vol. 76, 1979, pp. 106–110.
52. Lydia Villa-Komaroff, Argiris Efstratiadis, Stephanie Broome, Peter Lomedico, Richard Tizard, Stephen P. Naber, William L. Chick, and Walter Gilbert, "A Bacterial Clone Synthesizing Proinsulin," *Proc. Natl. Acad. Sci. USA*, vol. 75, 1978, 3727–3731; Annie C. Y. Chang, Jack H. Nunberg, Randal J. Kaufman, Henry A. Erlich, Robert T. Schimke, and Stanley N. Cohen, "Phenotypic Ex-

pression in *E. coli* of a DNA Sequence Coding for Mouse Dihydrofolate Reductase," *Nature*, vol. 275, 1978, pp. 617–624.

53. David V. Goeddel, Herbert L. Heynecker, Toyohara Hozumi, René Arentzen, Keiichi Itakura, Daniel G. Yansura, Michael J. Ross, Giuseppe Miozzari, Roberto Crea, and Peter Seeburg, "Direct Expression in *Escherichia coli* of a DNA Sequence Coding for Human Growth Hormone," *Nature*, vol. 281, 1979, pp. 544–548.
54. Shigekazu Nagata, Hideharu Taira, Alan Hall, Lorraine Johnsrud, Michel Streuli, Josef Ecsödi, Werner Boll, Kari Cantell, and Charles Weissmann, "Synthesis in *E. coli* of a Polypeptide with Human Leukocyte Interferon Activity," *Nature*, 1980, vol. 284, pp. 316–320; David V. Goeddel, Elizabeth Yelverton, Axel Ullrich, Herbert L. Heynecker, Giuseppe Miozzari, William Holmes, Peter H. Seeburg, Thomas Dull, Laurie May, Nowell Stebbing, Roberto Crea, Shuichiro Maeda, Russell McCandliss, Alan Sloma, John M. Tabor, Mitchell Gross, Philipp C. Familletti, and Sidney Pestka, "Human Leukocyte Interferon Produced by *E. coli* is Biologically Active," *Nature*, vol. 287, 1980, pp. 411–416.
55. Jean-Pierre Hernalsteens, Françoise Van Vliet, Marc De Beuckeleer, Ann Depicker, Gilbert Engler, Michel Lemmers, Marcelle Holsters, Marc Van Montagu, and Jeff Schell, "The Agrobacterium Tumefaciens Ti Plasmid as a Host Vector System for Introducing Foreign DNA in Plant Cells," *Nature*, vol. 287, 1980, pp. 654–656.
56. Jon W. Gordon, George A. Scangos, Diane J. Plotkin, James A. Barbosa, and Frank H. Ruddle, "Genetic Transformation of Mouse Embryos by Microinjection of Purified DNA," *Proc. Natl. Acad. Sci. USA*, vol. 77, 1980, pp. 7380–7384.
57. The experiment was a piece of classic molecular hybridization in solution. Its importance derived from the material used. Yuet Wai Kan, Mitchell S. Golbus, and Andrée M. Dozy, "Prenatal Diagnosis of α-Thalassemia: Clinical Application of Molecular Hybridization," *N. Engl. J. Med.*, vol. 295, pp. 1165–1167.

17. Split Genes and Splicing

1. A. S. Sarabhai, A. O. W. Stetton, Sydney Brenner, and A. Bolle, "Co-linearity of the Gene with the Polypeptide Chain," *Nature*, vol. 201, 1964, pp. 13–17; Charles Yanofsky, "Gene Structure and Protein Structure," *Harvey Lect.*, vol. 61, 1967, pp. 145–168.
2. S. M. Berget, A. J. Berk, T. Harrison, and P. A. Sharp, "Spliced Segments at the 5′ Termini of Adenovirus-2 Late mRNA: A Role for Heterogeneous Nuclear RNA in Mammalian Cells"; T. R. Broker, L. T. Chow, A. R. Dunn, R. E. Gelinas, J. A. Hassel, D. F. Klessig, J. B. Lewis, R. J. Roberts, and B. S. Zain, "Adenovirus-2 Messengers—An Example of Baroque Molecular Architecture";

and H. Westphal and S.-P. Lai, "Displacement Loops in Adenovirus DNA-RNA Hybrids," *Cold Spring Harbor Symp. Quant. Biol.*, vol. 42, Cold Spring Harbor, 1977, pp. 523–529, 531–553, and 555–558. Shortly afterwards, these results were published in scientific journals: Susan M. Berget, Claire Moore, and Phillip A. Sharp, "Spliced Segments at the $5'$ Terminus of Adenovirus-2 Late mRNA," *Proc. Natl. Acad. Sci. USA*, vol. 74, 1977, pp. 3171–3175; Louise T. Chow, Richard E. Gelinas, Thomas R. Broker, and Richard J. Roberts, "An Amazing Sequence Arrangement at the $5'$ Ends of Adenovirus 2 Messenger RNA," *Cell*, vol. 12, 1977, pp. 1–8; Daniel F. Klessig, "Two Adenovirus mRNAs Have a Common $5'$ Terminal Leader Sequence Encoded at Least 10 kb Upstream from Their Main Coding Regions," *Cell*, vol. 12, 1977, pp. 9–21; Ashley R. Dunn and John A. Hassell, "A Novel Method to Map Transcripts: Evidence for Homology between an Adenovirus mRNA and Discrete Multiple Regions of the Viral Genome," *Cell*, vol. 12, 1977, pp. 23–36; J. B. Lewis, C. W. Anderson, and J. F. Atkins, "Further Mapping of Late Adenovirus Genes by Cell-Free Translation of RNA Selected by Hybridization to Specific DNA Fragments," *Cell*, vol. 12, 1977, pp. 37–44.

3. B. G. Barrell, G. M. Air, and C. A. Hutchison III, "Overlapping Genes in Bacteriophage ϕX174," *Nature*, vol. 264, 1976, pp. 34–41.
4. Yosef Aloni, S. Bratosiw, Ravi Dhar, Orgad Laub, Mia Horowitz, and George Khoury, "Splicing of SV40 mRNAs: A Novel Mechanism for the Regulation of Gene Expression in Animal Cells," *Cold Spring Harbor Symp. Quant. Biol.*, vol. 42, Cold Spring Harbor, 1977, pp. 559–570; M.-T. Hsu and J. Ford, "A Novel Sequence Arrangement of SV40 late RNA," *Cold Spring Harbor Symp. Quant. Biol.*, vol. 42, Cold Spring Harbor, 1977, pp. 571–576; Yosef Aloni, Ravi Dhar, Orgad Laub, Mia Horowitz, and George Khoury, "Novel Mechanisms for RNA Maturation: The Leader Sequences of Simian Virus 40 mRNA Are Not Transcribed Adjacent to the Coding Sequences," *Proc. Natl. Acad. Sci. USA*, vol. 74, 1977, pp. 3686–3690.
5. A. J. Jeffreys and R. A. Flavell, "The Rabbit β-Globin Gene Contains a Large Insert in the Coding Sequence," *Cell*, vol. 12, 1977, pp. 1097–1108; R. Breathnach, J. L. Mandel, and P. Chambon, "Ovalbumin Gene Is Split in Chicken DNA," *Nature*, vol. 270, 1977, pp. 314–319; Shirley M. Tilghman, David C. Tiemeier, J. G. Seidman, B. Matija Peterlin, Margery Sullivan, Jacob V. Maizel, and Philip Leder, "Intervening Sequence of DNA Identified in the Structural Portion of a Mouse β-Globin Gene," *Proc. Natl. Acad. Sci. USA*, vol. 75, 1978, pp. 725–729; Christine Brack and Susumu Tonegawa, "Variable and Constant Parts of the Immunoglobulin Light Chain Gene of a Mouse Myeloma Cell are 1250 Non-Translated Bases Apart," *Proc. Natl. Acad. Sci. USA*, vol. 74, 1977, pp. 5652–5656.
6. Walter Gilbert, "Why Genes in Pieces," *Nature*, vol. 271, 1978, p. 501.

7. Francis Crick, "Split Genes and RNA Splicing," *Science,* vol. 204, 1979, pp. 264–271; R. Breathnach and P. Chambon, "Organization and Expression of Eucaryotic Split Genes Coding for Proteins," *Ann. Rev. Biochem.,* vol. 50, 1981, pp. 349–383.
8. J. L. Bos, C. Heyting, P. Borst, A. C. Arnberg, and E. F. J. Van Bruggen, "An Insert in the Single Gene for the Large Ribosomal RNA in Yeast Mitochondrial DNA," *Nature,* vol. 275, 1978, pp. 336–338.
9. Peter J. Curtis, Ned Mantei, Johan Van Den Berg, and Charles Weissmann, "Presence of a Putative 15S Precursor to β-Globin mRNA but not to α-Globin mRNA in Friend Cells," *Proc. Natl. Acad. Sci. USA,* vol. 74, 1977, pp. 3184–3188.
10. Gilbert, "Why Genes in Pieces"; Crick, "Split Genes and RNA Splicing," p. 264; Jan A. Witkowski, "The Discovery of 'Split' Genes: A Scientific Revolution," *TIBS,* vol. 13, 1988, pp. 110–113.
11. J. E. Darnell, L. Philipson, R. Wall, and M. Adesnik, "Polyadenylic Acid Sequences: Role in Conversion of Nuclear RNA into Messenger RNA," *Science,* vol. 174, 1971, pp. 507–510.
12. O. P. Samarina, "The Distribution and Properties of Cytoplasmic Deoxyribonucleic Acid–Like Ribonucleic Acid (Messenger Ribonucleic Acid)," *Biochem. Biophys. Acta,* vol. 91, 1964, pp. 688–691; G. P. Georgiev, "On the Structural Organization of Operon and the Regulation of RNA Synthesis in Animal Cells," *J. Theoret. Biol.,* vol. 25, 1969, pp. 473–490; G. P. Georgiev, A. P. Ryskov, C. Coutelle, V. L. Mantieva, and E. R. Avakyan, "On the Structure of Transcriptional Unit in Mammalian Cells," *BBA,* vol. 259, 1972, pp. 259–283; Robert A. Weinberg, "Nuclear RNA Metabolism," *Ann. Rev. Biochem.,* vol. 42, 1973, pp. 329–354.
13. Pierre Chambon, "Split Genes," *Scientific American,* vol. 244, May 1981, pp. 60–71.
14. Moshe Yaniv, "Génétique: le dogme de la colinéarité ébranlé," *La Recherche,* vol. 8, 1977, pp. 1100–1103.
15. Crick, "Split Genes and RNA Splicing," pp. 269–270.
16. Roy J. Britten and Eric H. Davidson, "Gene Regulation for Higher Cells: A Theory," *Science,* vol. 165, 1969, pp. 349–357.
17. Claude Jacq, Jaga Lazowska, and Piotr P. Slonimski, "Sur un nouveau mécanisme de la régulation de l'expression génétique," *Comptes rendus de l'Académie de sciences,* vol. 290, series D, 1980, pp. 89–92; Jaga Lazowska, Claude Jacq, and Piotr P. Slonimski, "Sequence of Introns and Flanking Exons in Wild-Type and Box3 Mutants of Cytochrome b Reveals an Interlaced Splicing Protein Coded by an Intron," *Cell,* vol. 22, 1980, pp. 333–348.
18. Piotr P. Slonimski, "Éléments hypothétiques de l'expression des gènes morcelés: protéines messagères de la membrane nucléaire," *Comptes rendus de l'Académie de sciences,* vol. 290, series D, 1980, pp. 331–334.

19. Antoine Danchin, "Règles de réécriture en biologie moléculaire," *Le Débat,* no. 3, July–August 1980, pp. 111–114.
20. Thomas R. Cech, Arthur J. Zaug, and Paula J. Grabowski, "*In vitro* Splicing of the Ribosomal RNA Precursor of Tetrahymena: Involvement of a Guanosine Nucleotide in the Excision of the Intervening Sequence," *Cell,* vol. 27, 1981, pp. 487–496; Kelly Kruger, Paula J. Grabowski, Arthur J. Zaug, Julie Sands, Daniel E. Gottschling, and Thomas R. Cech, "Self-Splicing RNA: Autoexcision and Autocyclization of the Ribosomal RNA Intervening Sequence of Tetrahymena," *Cell,* vol. 31, 1982, pp. 147–157.
21. Cecilia Guerrier-Takada, Katheleen Gardiner, Terry Marsh, Norman Pace, and Sydney Altman, "The RNA moiety of Ribonuclease P Is the Catalytic Subunit of the Enzyme," *Cell,* vol. 35, 1983, pp. 849–857.
22. Francis H. C. Crick, "The Origin of the Genetic Code," *J. Mol. Biol.,* vol. 38, 1968, pp. 367–379; Leslie E. Orgel, "Evolution of the Genetic Apparatus," *J. Mol. Biol.,* vol. 38, 1968, pp. 381–393.
23. The discovery of splicing confirmed Crick's intuition (Crick, "The Origin of the Genetic Code") that the primordial genetic material was RNA and not DNA: Darryl Reanney, "RNA Splicing and Polynucleotide Evolution," *Nature,* vol. 277, 1979, pp. 598–600. The discovery of self-splicing gave RNA the catalytic function it previously lacked: Walter Gilbert, "The RNA World," *Nature,* vol. 319, 1986, p. 618.
24. Gilbert, "Why Genes in Pieces"; James E. Darnell, Jr., "Implication of RNA: RNA Splicing in Evolution of Eukaryotic Cells," *Science,* vol. 202, 1978, pp. 1257–1260.
25. Gilbert, "Why Genes in Pieces."
26. Charles S. Craik, Stephen Sprang, Robert Fletterick, and William J. Rutter, "Intron-Exon Splice Junctions Map at Protein Surfaces," *Nature,* vol. 299, 1982, pp. 180–182; Charles S. Craik, William J. Rutter, and Robert Fletterick, "Splice Junctions: Association with Variation in Protein Structure," *Science,* vol. 220, 1983, pp. 1125–1129.
27. Colin C. F. Blake, "Do Genes-in-Pieces Imply Proteins-in-Pieces?" *Nature,* vol. 273, 1978, p. 267.
28. Walter Gilbert, "Why Genes in Pieces."
29. S. Ohno, *Evolution by Gene Duplication,* Springer-Verlag, New York, 1970.
30. Chambon, "Split Genes."
31. Darnell, "Implication of RNA:RNA Splicing"; Susumu Tonegawa, Allan M. Maxam, Richard Tizard, Ora Bernard, and Walter Gilbert, "Sequence of a Mouse Germ-Line Gene for a Variable Region of an Immunoglobin Light Chain," *Proc. Natl. Acad. Sci. USA,* vol. 75, 1978, pp. 1485–1489; Gilbert, "Why Genes in Pieces."
32. The concept of tinkering—introduced by Jacob in 1977, before the existence of split genes was known—found an excellent application in this new con-

text: F. Jacob, "Evolution and Tinkering," *Science*, vol. 196, 1977, pp. 1161–1166; François Jacob, *Le Jeu des possibles*, Fayard, Paris, 1981 (translated as *The Possible and the Actual*, University of Washington Press, Seattle, 1982).

33. Carmen Quinto, Margarita Quiroga, William F. Swain, William C. Nikovits, Jr., David N. Standring, Raymond L. Pictet, Pablo Valenzuela, and William J. Rutter, "Rat Preprocarboxypeptidase A: cDNA Sequence and Preliminary Characterization of the Gene," *Proc. Natl. Acad. Sci. USA*, vol. 79, 1982, pp. 31–35.
34. Margaret Leicht, George L. Long, T. Chandra, Kotoku Kurachi, Vincent J. Kidd, Myles Mace, Jr., Earl W. Davies, and Savio L. C. Woo, "Sequence Homology and Structural Comparison between the Chromosomal Human α1-Antitrypsin and Chicken Ovalbumin Genes," *Nature*, vol. 297, 1982, pp. 655–659.
35. W. Ford Doolittle, "Genes in Pieces: Were They Ever Together?" *Nature*, vol. 272, 1978, pp. 581–582; Darnell, "Implication of RNA:RNA Splicing."
36. Crick, "Split Genes and RNA Splicing," p. 269.
37. Claude Jacq, J. R. Miller, and G. G. Brownlee, "A Pseudogene Structure in 5S DNA of *Xenopus Laevis*," *Cell*, vol. 12, 1977, pp. 109–120; Y. Nishioka, A. Leder, and P. Leder, "Unusual α-Globin-Like Gene That Has Cleanly Lost Both Globin Intervening Sequences," *Proc. Natl. Acad. Sci. USA*, vol. 77, 1980, pp. 2806–2809.
38. B. G. Barrell, A. T. Bankier, and J. Drouin, "A Different Genetic Code in Human Mitochondria," *Nature*, vol. 282, 1979, pp. 189–194.
39. Stuart Horowitz and Martin A. Gorovsky, "An Unusual Genetic Code in Nuclear Genes of Tetrahymena," *Proc. Natl. Acad. Sci. USA*, vol. 82, 1985, pp. 2452–2455.
40. François Caron and Eric Meyer, "Does *Paramecium primaurelia* Use a Different Genetic Code in Its Macronucleus?" *Nature*, vol. 314, 1985, pp. 185–188; J. R. Preer, L. B. Preer, B. M. Rudman, and A. J. Barnett, "Deviation from the Universal Code Shown by the Gene for Surface Protein 51.A in *Paramecium*," *Nature*, vol. 314, 1985, pp. 188–190.
41. Rob Benne, Janny Van den Burg, Just P. J. Brakenhoff, Paul Sloof, Jacques H. Van Boom, and Marike C. Tromp, "Major Transcript of the Frameshifted Cox II Gene from Trypanosome Mitochondria Contains Four Nucleotides That Are Not Encoded in the DNA," *Cell*, vol. 46, 1986, pp. 819–826.
42. Beat Blum, Nancy R. Sturm, Agda M. Simpson, and Larry Simpson, "Chimeric gRNA-mRNA Molecules with Oligo (U) Tails Covalently Linked at Sites of RNA Editing Suggest that U Addition Occurs by Transesterification," *Cell*, vol. 65, 1991, pp. 543–550.
43. Crick, "On Protein Synthesis."
44. Joshua Lederberg, "Genes and Antibodies," *Science*, vol. 129, 1959, pp. 1649–1653.
45. Frank MacFarlane Burnet, *The Clonal Selection Theory of Acquired Immunity*, Cambridge University Press, Cambridge, England, 1959. This model had first been proposed in 1957 in an Australian journal: Frank MacFarlane Burnet, "A

Modification of Jerne's Theory of Antibody Production Using the Concept of Clonal Selection," *Australian Journal of Science,* vol. 20, 1957, pp. 67–69.

46. G. J. V. Nossal and Joshua Lederberg, "Antibody Production by Single Cells," *Nature,* vol. 181, 1958, pp. 1419–1420.
47. Lederberg, "Genes and Antibodies," p. 1649.
48. The adoption of Burnet's theory was accompanied by a profound transformation of immunology and led to the concept of the immune system: Anne-Marie Moulin, "De l'analyse au système: le développement de l'immunologie," *Rev. Hist. Sci.,* vol. 36, 1983, pp. 49–67; Anne-Marie Moulin, *Le Dernier Langage de la médecine,* second part.
49. Nobumichi Hozumi and Susumu Tonegawa, "Evidence for Somatic Rearrangement of Immunoglobulin Genes Coding for Variable and Constant Regions," *Proc. Natl. Acad. Sci. USA,* vol. 73, 1976, pp. 3628–3632.
50. Stephen M. Hedrick, David I. Cohen, Ellen A. Nielsen, and Mark M. Davis, "Isolation of cDNA Clones Encoding T Cell-Specific Membrane-Associated Proteins," *Nature,* vol. 308, 1984, pp. 149–153.

18. A New Molecular Biology

1. Michel Morange, *Une lecture du vivant: Histoire et épistémologie de la biologie moléculaire,* CIACO, Louvain-la-Neuve, 1986.
2. Niels Bohr, "Light and Life," *Nature,* vol. 131, 1933, pp. 421–423 and 457–459.

19. The Discovery of Oncogenes

1. Michel Morange, "The Discovery of Cellular Oncogenes," *Hist. Life Phil. Sci.,* vol. 15, 1993, pp. 45–59; "From the Regulatory Vision of Cancer to the Oncogene Paradigm," *J. Hist. Biol.,* vol. 30, 1997, pp. 1–27; Natalie Angier, *Natural Obsessions: The Search for the Oncogene,* Houghton Mifflin Company, Boston, 1988. The work of Joan Fujimura casts light on the "strategic" stakes involved in the discovery of oncogenes: Joan Fujimura, "Constructing Doable Problems in Cancer Research: Articulating Alignments," *Soc. Stud. Sci.,* vol. 17, 1987, pp. 257–293; "The Molecular Biological Bandwagon in Cancer Research: Where Social Worlds Meet," *Social Problems,* vol. 35, 1988, pp. 261–283; *Crafting Science: A Sociohistory of the Quest for the Genetics of Cancer,* Harvard University Press, Cambridge, Mass., 1996. A very simplified presentation of this history can be found in Harold Varmus and Robert A. Weinberg, *Genes and the Biology of Cancer,* Scientific American Library, New York, 1993.

2. Robert J. Huebner and George J. Todaro, "Oncogenes of RNA Tumor Viruses as Determinants of Cancer," *Proc. Natl. Acad. Sci. USA*, vol. 64, 1969, pp. 1087–1094. The work of Dulbecco's group on oncogenic DNA viruses, the polyoma virus and SV40, showed that transformation was due to a gene carried by the virus. But the mechanism of action of this gene remained unknown. Furthermore, Dulbecco thought that transformation was the result of other cellular changes that took place after viral invasion. Therefore his findings were not incompatible with the model of Huebner and Todaro. See Renato Dulbecco, "From the Molecular Biology of Oncogenic DNA Viruses to Cancer," *Science*, vol. 192, 1976, pp. 437–440. Research on oncogenic viruses received significant funding, enjoying an important position in the crusade against cancer launched in the United States in the late 1960s. It was hoped that these studies would lead to the rapid development of diagnostic tools and therapeutic methods, benefiting both patients and the pharmaceutical industry.
3. Howard M. Temin, "The Protovirus Hypothesis," *J. Natl. Cancer Inst.*, vol. 46, 1971, pp. iii–viii.
4. Edward M. Scolnick, Elaine Rands, David Williams, and Wade P. Parks, "Studies on the Nucleic Acid Sequences of Kirsten Sarcoma Virus: A Model for Formation of a Mammalian RNA-Containing Sarcoma Virus," *J. Virol.*, vol. 12, 1973, pp. 458–463.
5. François Jacob, "Comments," *Cancer Research*, vol. 20, 1960, pp. 695–697. The analogy between lysogeny and cancer had already been emphasized by André Lwoff seven years earlier: André Lwoff, "Lysogeny," *Bacteriol. Rev.*, vol. 17, 1953, pp. 269–337. See also Charles Galperin, "Virus, provirus et cancer," *Rev. Hist. Sci.*, vol. 47 (1), 1994, pp. 7–56.
6. See Chapter 5.
7. Howard M. Temin, "On the Origin of the Genes for Neoplasia: G. H. A. Clowes Memorial Lectures," *Cancer Research*, vol. 34, 1974, pp. 2835–2841.
8. Dominique Stehelin, Ramareddy V. Guntaka, Harold E. Varmus, and J. Michael Bishop, "Purification of DNA Complementary to Nucleotide Sequences Required for Neoplastic Transformation of Fibroblasts by Avian Sarcoma Viruses," *J. Mol. Biol.*, vol. 101, 1975, pp. 349–365; Dominique Stehelin, Harold E. Varmus, J. Michael Bishop, and Peter K. Vogt, "DNA Related to the Transforming Gene(s) of Avian Sarcoma Viruses Is Present in Normal Avian DNA," *Nature*, vol. 260, 1976, pp. 170–173.
9. Deborah H. Spector, Harold E. Varmus, and J. Michael Bishop, "Nucleotide Sequences Related to the Transforming Gene of Avian Sarcoma Virus Are Present in DNA of Uninfected Vertebrates," *Proc. Natl. Acad. Sci. USA*, vol. 75, 1978, pp. 4102–4106.
10. See, for example, J. Michael Bishop, "Enemies Within: The Genesis of Retrovirus Oncogenes," *Cell*, vol. 23, 1981, pp. 5–6.
11. Angier, *Natural Obsessions.*

12. Deborah H. Spector, Karen Smith, Thomas Padgett, Pamela McCombe, Daisy Roulland-Dussoix, Carlo Moscovici, Harold E. Varmus, and J. Michael Bishop, "Uninfected Avian Cells Contain RNA Related to the Transforming Gene of Avian Sarcoma Viruses," *Cell*, vol. 13, 1978, pp. 371–379; Deborah H. Spector, Barbara Baker, Harold E. Varmus, and J. Michael Bishop, "Characteristics of Cellular RNA Related to the Transforming Gene of Avian Sarcoma Viruses," *Cell*, vol. 13, 1978, pp. 381–386.
13. Michel Morange, "From the Regulatory Vision of Cancer to the Oncogene Paradigm," *J. Hist. Biol.*, vol. 30, 1997, pp. 1–27.
14. In 1973 Graham and Van der Eb had developed a technique for "transfecting" cells with exogenous DNA. In 1979 this technique became operational through cotransfection with a resistance gene (see Chapter 16); Weinberg and Cooper carried out their transfection experiments in the same year. In these experiments, the oncogene is positively selected—transformed cells grow quicker and can easily be detected and isolated in the culture dish.
15. Joyce McCann, Edmund Choi, Edith Yamasaki, and Bruce N. Ames, "Detection of Carcinogens as Mutagens in the *Salmonella*/Microsome Test: Assay of 300 Chemicals," *Proc. Natl. Acad. Sci. USA*, vol. 72, 1975, pp. 5135–5139.
16. Chiaho Shih, Ben-Zion Shilo, Mitchell P. Goldfarb, Ann Dannenberg, and Robert A. Weinberg, "Passages of Phenotypes of Chemically Transformed Cells via Transfection of DNA and Chromatin," *Proc. Natl. Acad. Sci. USA*, vol. 76, 1979, pp. 5714–5718; Geoffrey M. Cooper, Sharon Okenquist, and Lauren Silverman, "Transforming Activity of DNA of Chemically Transformed and Normal Cells," *Nature*, vol. 284, 1980, pp. 418–421.
17. This last result was obtained only by Weinberg's group. Cooper showed that oncogenes could be extracted from normal, untransformed cells by cleaving DNA into sufficiently small fragments. Cooper's results agreed with the then-dominant conception of cancer as deregulation. Michel Morange, "From the Regulatory Vision of Cancer to the Oncogene Paradigm," *J. Hist. Biol.*, vol. 30, 1997, pp. 1–27.
18. Luis F. Parada, Clifford J. Tabin, Chiaho Shih, and Robert A. Weinberg, "Human E. J. Bladder Carcinoma Oncogene is Homologue of Harvey Sarcoma Virus *Ras* Gene," *Nature*, vol. 297, 1982, pp. 474–478; Channing J. Der, Theodore G. Krontiris, and Geoffrey M. Cooper, "Transforming Genes of Human Bladder and Lung Carcinoma Cell Lines Are Homologous to the *Ras* Genes of Harvey and Kirsten Sarcoma Viruses," *Proc. Natl. Acad. Sci. USA*, vol. 79, 1982, pp. 3637–3640.
19. Clifford J. Tabin, Scott M. Bradley, Cornelia I. Bargmann, Robert A. Weinberg, Alex G. Papageorge, Edward M. Scolnick, Ravi Dhar, Douglas R. Lowy, and Esther H. Chang, "Mechanism of Activation of a Human Oncogene," *Nature*, vol. 300, 1982, pp. 143–149; E. Premkumar Reddy, Roberta K. Reynolds, Eugenio Santos, and Mariano Barbacid, "A Point Mutation Is Responsible for

the Acquisition of Transforming Properties by the T24 Human Bladder Carcinoma Oncogene," *Nature,* vol. 300, 1982, pp. 149–152.

20. William S. Hayward, Benjamin G. Neel, and Susan M. Astrin, "Activation of a Cellular *onc* Gene by Promoter Insertion in ALV-Induced Lymphoid Leukosis," *Nature,* vol. 290, 1981, pp. 475–480; Steven Collins and Mark Groudine, "Amplification of Endogenous *Myc*-Related DNA Sequences in a Human Myeloid Leukaemia Cell Line," *Nature,* vol. 298, 1982, pp. 679–681; Philip Leder, Jim Battey, Gilbert Lenoir, Christopher Moulding, William Murphy, Huntington Potter, Timothy Stewart, and Rebecca Taub, "Translocations among Antibody Genes in Human Cancer," *Science,* vol. 222, 1983, pp. 765–771.
21. Russell F. Doolittle, Michael W. Hunkapiller, Leroy E. Hood, Sushilkumar G. Devare, Keith C. Robbins, Stuart A. Aaronson, and Harry N. Antoniades, "Simian Sarcoma Virus *Onc* Gene, v-sis, Is Derived from the Gene (or Genes) Encoding a Platelet-Derived Growth Factor," *Science,* vol. 221, 1983, pp. 275–277; Michael D. Waterfield, Geoffrey T. Scrace, Nigel Whittle, Paul Sroobant, Ann Johnsson, Åke Wasteson, Bengt Westermark, Carl-Henrik Heldin, Jung San Huang, and Thomas F. Deuel, "Platelet-Derived Growth Factor Is Structurally Related to the Putative Transforming Protein $p28^{sis}$ of Simian Sarcoma Virus," *Nature,* vol. 304, 1983, pp. 35–39.
22. J. Downward, Y. Yarden, E. Mayes, G. Scrace, N. Totty, P. Stockwell, A. Ullrich, J. Schlessinger, and M. D. Waterfield, "Close Similarity of Epidermal Growth Factor Receptor and v-*erb*-B Oncogene Protein Sequences," *Nature,* vol. 307, 1984, pp. 521–527.
23. James B. Hurley, Melvin I. Simon, David B. Teplow, Janet D. Robishaw, and Alfred G. Gilman, "Homologies between Signal Transducing G Proteins and *ras* Gene Products," *Science,* vol. 226, 1984, pp. 860–862.
24. Kathleen Kelly, Brent H. Cochran, Charles D. Stiles, and Philip Leder, "Cell-Specific Regulation of the *c-myc* Gene by Lymphocyte Mitogens and Platelet-Derived Growth Factor," *Cell,* vol. 35, 1983, pp. 603–610; Wiebe Kruijer, Jonathan A. Cooper, Tony Hunter, and Inder M. Verma, "Platelet-Derived Growth Factor Induces Rapid but Transient Expression of the c-*fos* Gene and Protein," *Nature,* vol. 312, 1984, pp. 711–716; Rolf Müller, Rodrigo Bravo and Jean Burckhardt, Tom Curran, "Induction of c-*fos* Gene and Protein by Growth Factors Precedes Activation of c-*myc,*" *Nature,* vol. 312, 1984, pp. 716–720.
25. D. Defeo-Jones, E. M. Scolnick, R. Koller, and R. Dhar, "ras-Related Gene Sequences Identified and Isolated from *Saccharomyces cerevisiae,*" *Nature,* vol. 306, 1983, pp. 707–709.
26. Michael J. Berridge and Robin F. Irvine, "Inositol Triphosphate, a Novel Second Messenger in Cellular Signal Transduction," *Nature,* vol. 312, 1984, pp. 315–321; Yasutomi Nishizuka, "The Role of Protein Kinase C in Cell Surface Signal Transduction and Tumour Promotion," *Nature,* vol. 308, 1984, pp. 693–698.
27. Minoo Rassoulzadegan, Alison Cowie, Antony Carr, Nicolas Glaichenhaus,

Robert Kamen, and François Cuzin, "The Roles of Individual Polyoma Virus Early Proteins in Oncogenic Transformation," *Nature*, vol. 300, 1982, pp. 713–718.

28. Harmut Land, Luis F. Parada, and Robert A. Weinberg, "Cellular Oncogenes and Multistep Carcinogenesis," *Science*, vol. 222, 1983, pp. 771–778.
29. Robert A. Weinberg, "The Action of Oncogenes in the Cytoplasm and Nucleus," *Science*, vol. 230, 1985, pp. 770–776.
30. Alfred G. Knudson, Jr., "Mutation and Cancer: Statistical Study of Retinoblastoma," *Proc. Natl. Acad. Sci. USA*, vol. 68, 1971, pp. 820–823.
31. Stephen H. Friend, René Bernards, Snezna Rogelj, Robert A. Weinberg, Joyce M. Rapaport, Daniel M. Albert, and Thaddeus P. Dryja, "A Human DNA Segment with Properties of the Gene That Predisposes to Retinoblastoma and Osteosarcoma," *Nature*, vol. 323, 1986, pp. 643–646.
32. Fujimura, "The Molecular Biological Bandwagon in Cancer Research," pp. 261–283.
33. George Klein, "The Role of Gene Dosage and Genetic Transposition in Carcinogenesis," *Nature*, vol. 294, 1981, pp. 313–318.
34. Jean-Paul Gaudillière, "Oncogenes as Metaphors for Human Cancer: Articulating Laboratory Practices and Medical Demands," in Ilana Lowy (ed.), *Medicine and Change: Historical and Sociological Studies of Medical Innovation*, INSERM, J. Libbey Eurotext, 1992, pp. 213–247.
35. Peter H. Duesberg, "Cancer Genes: Rare Recombinants Instead of Activated Oncogenes (a Review)," *Proc. Natl. Acad. Sci. USA*, vol. 84, 1987, pp. 2117–2124.

20. From DNA Polymerase to the Amplification of DNA

1. Kary B. Mullis, "The Unusual Origin of the Polymerase Chain Reaction," *Scientific American*, vol. 262, April 1990, pp. 36–43; Paul Rabinow, *Making PCR: A Story of Biotechnology*, University of Chicago Press, Chicago, 1996.
2. Henry A. Erlich, David Gelfand, and John J. Sninsky, "Recent Advances in the Polymerase Chain Reaction," *Science*, vol. 252, 1991, pp. 1643–1651.
3. Horace F. Judson, *The Eighth Day of Creation: The Makers of the Revolution in Biology*, Simon and Schuster, New York, 1979, p. 322.
4. Franklin H. Portugal and Jack S. Cohen, *A Century of DNA: A History of the Discovery of the Structure and Function of the Genetic Substance*, MIT Press, Cambridge, Mass., 1977, pp. 314–317.
5. Arthur Kornberg, *For the Love of Enzymes: The Odyssey of a Biochemist*, Harvard University Press, Cambridge, Mass., 1989; Arthur Kornberg, "Never a Dull Enzyme," *Ann. Rev. Biochem.*, vol. 58, 1989, pp. 1–30. Two reviews of Kornberg's autobiographical account are also worth reading: Pnina G. Abir-

Am, "Noblesse Oblige: Lives of Molecular Biologists," *Isis*, vol. 82, 1991, pp. 326–343; Jan Sapp, "Portraying Molecular Biology," *J. Hist. Biol.*, vol. 25, 1992, pp. 149–155.

6. Kornberg, "Never a Dull Enzyme," p. 6.
7. Kornberg, *For the Love of Enzymes*, pp. 121–122.
8. Kornberg, "Never a Dull Enzyme," p. 11.
9. Marianne Grunberg-Manago and Severo Ochoa, "Enzymatic Synthesis and Breakdown of Polynucleotides: Polynucleotide Phosphorylase," *JACS*, vol. 77, 1955, pp. 3165–3166.
10. Uriel Z. Littauer and Arthur Kornberg, "Reversible Synthesis of Polyribonucleotides with an Enzyme from *Escherichia coli*," *J. Biol. Chem.*, vol. 226, 1957, pp. 1077–1092.
11. Arthur Kornberg, "Pathways of Enzymatic Synthesis of Nucleotides and Polynucleotides," in W. D. McElroy and B. Glass (eds.), *The Chemical Basis of Heredity*, Mac Collum-Pratt Symposium, Johns Hopkins University Press, Baltimore, pp. 579–608; I. R. Lehman, Maurice J. Bessman, Ernest S. Simms, and Arthur Kornberg, "Enzymatic Synthesis of Deoxyribonucleic Acid. I. Preparation of Substrates and Partial Purification of an Enzyme from *Escherichia coli*," *J. Biol. Chem.*, vol. 233, 1958, pp. 163–170; Maurice J. Bessman, I. R. Lehman, Ernest S. Simms, and Arthur Kornberg, "Enzymatic Synthesis of Deoxyribonucleic Acid. II. General Properties of the Reaction," *J. Biol. Chem.*, vol. 233, 1958, pp. 171–177.
12. I. R. Lehman, Steven R. Zimmerman, Julius Adler, Maurice J. Bessman, Ernest S. Simms, and Arthur Kornberg, "Enzymatic Synthesis of Deoxyribonucleic Acid. V. Chemical Composition of Enzymatically Synthesized Deoxyribonucleic Acid," *Proc. Natl. Acad. Sci. USA*, vol. 44, 1958, pp. 1191–1196.
13. Arthur Kornberg, "Biologic Synthesis of Deoxyribonucleic Acid: An Isolated Enzyme Catalyzes Synthesis of This Nucleic Acid in Response to Directions from Pre-existing DNA," *Science*, vol. 131, 1960, pp. 1503–1508.
14. This view of enzymes as demiurges and the ease with which it was accepted by molecular biologists have been discussed by René Thom: "I am surprised to see how . . . biologists react to the questions of molecular biology. The behavior of macromolecules is something extraordinarily surprising, and yet in the literature biologists seem to find it quite natural. In DNA replication, in the manner in which the helix splits and the two fragments separate into two distinct cells, they see only the work of enzymes, which they think explains everything" (René Thom, *Paraboles et catastrophes: entretiens sur les mathématiques, la science et la philosophie*, Flammarion, Paris, 1983, p. 131).
15. Kornberg, "Never a Dull Enzyme," p. 13; Kornberg, *For the Love of Enzymes*, p. 163.
16. Mehran Goulian and Arthur Kornberg, "Enzymatic Synthesis of DNA, XXIII. Synthesis of Circular Replicative Form of Phage ϕX174 DNA," *Proc. Natl.*

Acad. Sci. USA, vol. 58, 1967, pp. 1723–1730; Mehran Goulian, Arthur Kornberg, and Robert Sinsheimer, "Enzymatic Synthesis of DNA, XXIV. Synthesis of Infectious Phage ϕX174 DNA," *Proc. Natl. Acad. Sci. USA*, vol. 58, 1967, pp. 2321–2328.

17. Kornberg, "Never a Dull Enzyme," p. 14. This experiment, however, had been preceded two years earlier by a similar experiment carried out on an RNA virus. (S. Spiegelman, T. Haruna, I. B. Holland, G. Beaudreau, and D. Mills, "The Synthesis of a Self-Propagating and Infectious Nucleic Acid with a Purified Enzyme," *Proc. Natl. Acad. Sci. USA*, vol. 54, 1965, pp. 919–927).
18. Paula de Lucia and John Cairns, "Isolation of an *E. coli* Strain with a Mutation Affecting DNA Polymerase," *Nature*, vol. 224, 1969, pp. 1164–1166.
19. "How Relevant Is Kornberg Polymerase?" *Nature New Biology*, vol. 229, 1971, pp. 65–66; "Is Kornberg Junior Enzyme the True Replicase?" *Nature New Biology*, vol. 230, 1971, p. 258.
20. This "nick translation" activity of DNA Polymerase I, detected by Kornberg's group in 1970, was used by many laboratories to label DNA molecules radioactively. A complete description of this technique can be found in Peter W. J. Rigby, Marianne Dieckmann, Carl Rhodes, and Paul Berg, "Labelling Deoxyribonucleic Acid to High Specific Activity *in Vitro* by Nick Translation with DNA Polymerase I," *J. Mol. Biol.*, vol. 113, 1977, pp. 237–251.
21. Frederick Sanger, "Sequences, Sequences and Sequences," *Ann. Rev. Biochem.*, vol. 57, 1988, pp. 1–28.
22. F. Sanger, S. Nicklen, and A. R. Coulson, "DNA Sequencing with Chain Terminating Inhibitors," *Proc. Natl. Acad. Sci. USA*, vol. 74, 1977, pp. 5463–5467.
23. Allan M. Maxam and Walter Gilbert, "A New Method for Sequencing DNA," *Proc. Natl. Acad. Sci. USA*, vol. 74, 1977, pp. 560–564.
24. Mullis, "The Unusual Origin of the Polymerase Chain Reaction."
25. Randall K. Saiki, Stephen Scharf, Fred Faloona, Kary B. Mullis, Glenn T. Horn, Henry A. Erlich, and Norman Arnheim, "Enzymatic Amplification of β-Globin Genomic Sequences and Restriction Site Analysis for Diagnosis of Sickle Cell Anemia," *Science*, vol. 230, 1985, pp. 1350–1354.
26. Randall K. Saiki, David H. Gelfand, Susanne Stoffel, Stephen J. Scharf, Russell Higuchi, Glenn T. Horn, Kary B. Mullis, and Henry A. Erlich, "Primer-Directed Enzymatic Amplification of DNA with a Thermostable DNA Polymerase," *Science*, vol. 239, 1988, pp. 487–491.
27. Erlich, Gelfand, and Sninsky, "Recent Advances in the Polymerase Chain Reaction," pp. 1643–1651.
28. According to the definition given by Harriet Zuckerman and Joshua Lederberg, "Post-Mature Scientific Discovery?" *Nature*, vol. 324, 1986, pp. 629–631.
29. Mullis, "The Unusual Origin of the Polymerase Chain Reaction," p. 43. On the basis of these arguments, Du Pont de Nemours began proceedings to annul

the patent held by Cetus: Marcia Barinaga, "Biotech Nightmare: Does Cetus Own PCR?" *Science*, vol. 251, 1991, pp. 739–740.

30. K. Kleppe, E. Ohtsuka, R. Kleppe, I. Molineux, and H. G. Khorana, "Studies on Polynucleotides XCVI. Repair Replication of Short Synthetic DNAs as Catalyzed by DNA Polymerase," *J. Mol. Biol.*, vol. 56, 1971, pp. 341–361.
31. Erlich, Gelfand, and Sninsky, "Recent Advances in the Polymerase Chain Reaction," p. 1650.
32. See also the allusion to "deoxyribonucleic bombs," which "exploded" in Kary Mullis's head after his discovery: Mullis, "The Unusual Origin of the Polymerase Chain Reaction," p. 41.
33. Kimberley Carr, "Nobel Rewards Two Laboratory Revolutions," *Nature*, vol. 365, 1993, p. 685; Tim Appenzeller, "Chemistry: Laurels for a Late-Night Brainstorm," *Science*, vol. 262, 1993, pp. 506–507. The Nobel Prize was jointly awarded to Michael Smith for developing the technique of directed mutagenesis (see Chapter 16).

21. Molecular Biology in the Life Sciences

1. This contemporary question should be distinguished from the historical question of the role of reductionism in the birth of molecular biology. I have described the reductionist view of biological phenomena that was held by the members of the Rockefeller Institute, and also the antireductionist approach of Bohr and Max Delbrück (Chapters 4, 7, 8, 9). The positive role of reductionist philosophy in the birth of molecular biology has been emphasized by Fuerst (John A. Fuerst, "The Role of Reductionism in the Development of Molecular Biology: Peripheral or Central," *Soc. Stud. Sci.*, vol. 12, 1982, pp. 241–278). By contrast, Schaffner has argued that reductionism played a minor role in the development of molecular biology (Kenneth A. Schaffner, "The Peripherality of Reductionism in the Development of Molecular Biology," *J. Hist. Biol.*, vol. 7, 1974, pp. 111–139).
2. Michael Ruse, "Molecular Revolution in Genetics," in *Is Science Sexist?* D. Reidel Publishing Company, Dordrecht, Holland, 1981. These readjustments became increasingly difficult with molecular biologists' "deconstruction" of the concept of the gene (see Chapter 17). See Raphael Falk, "What Is a Gene?" *Stud. Hist. Phil. Sci.*, vol. 17, 1986, pp. 133–173; Petter Portin, "The Concept of the Gene: Short History and Present Status," *Q. Rev. Biol.*, vol. 68, 1993, pp. 173–223.
3. The American philosopher of science Harold Kincaid chose examples from cell biology to show that biology has not been reduced to biochemistry (molecular biology, according to the definition given here), and that instead vari-

ous disciplines cooperated to explain life (Harold Kincaid, "Molecular Biology and the Unity of Science," *Philosophy of Science,* vol. 57, 1990, pp. 575–593). Many other examples can be found in immunology or, as shown, oncology.

4. See, for example, Francisco J. Ayala, "Thermodynamics, Information and Evolution: The Problem of Reductionism," *Hist. Phil. Life Sci.,* vol. 11, 1989, pp. 115–120. Discussed by Ernst Mayr, *The Growth of Biological Thought: Diversity, Evolution, and Inheritance,* Harvard University Press, Cambridge, Mass., 1982.
5. Kincaid, "Molecular Biology and the Unity of Science."
6. Steven Rose, "Reflections on Reductionism," *TIBS,* vol. 13, 1988, pp. 160–162. It should be pointed out that the experimental reductionism of molecular biologists is relatively limited. They only rarely "descend" to the physico-chemical level and often consider macromolecules simply black boxes whose molecular complexity is only moderately interesting. See Morange, *Une lecture du vivant,* pp. 69–108. Moreover, for molecular biologists, the present structure of organisms is the result of their evolutionary history. Physico-chemical rationality is not sufficient to explain the molecular mechanisms that are now known to exist.
7. More than twenty years ago Steven Rose called for this kind of cooperation between the various disciplines that study the brain: Steven Rose, *The Conscious Brain,* Random House, 1976.
8. Kincaid, "Molecular Biology and the Unity of Science."
9. The desire to place molecular biology and the theory of evolution on an equal footing is particularly clear in a 1959 article by Ernst Mayr: "Where Are We?" *Cold Spring Harbor Symp. Quant. Biol.,* vol. 24, 1959, pp. 1–14.
10. Ernst Mayr and William B. Provine (eds.), *The Evolutionary Synthesis: Perspectives on the Unification of Biology,* Harvard University Press, Cambridge, Mass., 1980; B. Smocovitis, "Unifying Biology: The Evolutionary Synthesis and Evolutionary Biology," *J. Hist. Biol.,* vol. 25, 1992, pp. 1–65. The demand for autonomy for biology is particularly strong in Ernst Mayr's writings (Mayr, *The Growth of Biological Thought).*
11. George G. Simpson, "Biology and the Nature of Science: Unification of the Sciences Can Be Most Meaningfully Sought through Study of the Phenomena of Life," *Science,* vol. 139, 1963, pp. 81–88.
12. Walter M. Fitch, "The Challenges to Darwinism since the Last Centennial and the Impact of Molecular Studies," *Evolution,* vol. 36, 1982, pp. 1133–1143. The almost complete universality of the genetic code strikingly confirmed that all living beings are descended from the same ancestral organism.
13. Richard C. Lewontin, *The Genetic Basis of Evolutionary Change,* Columbia University Press, New York, 1974; Jean Gayon, *Darwin et l'après-Darwin: une histoire de l'hypothèse de sélection naturelle,* Kimé, Paris, 1992, pp. 390–405 (translated as *Darwinism's Struggle for Survival: Heredity and the Hypothesis*

of Natural Selection, Cambridge University Press, Cambridge, England, forthcoming).

14. Motoo Kimura, "Evolutionary Rate at the Molecular Level," *Nature,* vol. 217, 1968, pp. 624–626; Jack L. King and Thomas H. Jukes, "Non Darwinian Evolution: Random Fixation of Selectively Neutral Mutations," *Science,* vol. 164, 1969, pp. 788–798; Michael R. Dietrich, "The Origins of the Neutral Theory of Molecular Evolution," *J. Hist. Biol.,* vol. 27, 1994, pp. 21–59.
15. Lynn Margulis, *Symbiosis in Cell Evolution,* W. H. Freeman and Co., San Francisco, 1981; W. Fred Doolittle, "Revolutionary Concepts in Evolutionary Cell Biology," *TIBS,* vol. 5, 1980, pp. 146–149.
16. Willi Hennig, *Grundzüge einer Theorie der phylogenetischen Systematik,* Deutscher Zentralverlag, Berlin, 1950; republished as *Phylogenetic Systematics,* University of Illinois Press, Urbana, 1966.
17. Rebecca L. Cann, Mark Stoneking, and Allan C. Wilson, "Mitochondrial DNA and Human Evolution," *Nature,* vol. 325, 1987, pp. 31–36.
18. Edward O. Wilson, *Sociobiology: The New Synthesis,* the Belknap Press of Harvard University Press, Cambridge, Mass., 1975.
19. Stephen J. Gould and Niles Eldredge, "Punctuated Equilibria: The Tempo and Mode of Evolution Reconsidered," *Paleobiology,* vol. 3, 1977, pp. 115–151. An analysis of this crisis of neo-Darwinism can be found in David Collingridge and Mark Earthy, "Science under Stress: Crisis in Neo-Darwinism," *Hist. Phil. Life Sci.,* vol. 12, 1990, pp. 3–26.
20. Peter G. Williamson, "Palaeontological Documentation of Speciation in Cenozoic Molluscs from Turkana Basin," *Nature,* vol. 293, 1981, pp. 437–443.
21. Pere Alberch, "Ontogenesis and Morphological Diversification," *Amer. Zool.,* vol. 20, 1980, pp. 653–667; Stephen Jay Gould, "Darwinism and the Expansion of Evolutionary Theory," *Science,* vol. 216, 1982, pp. 380–387.
22. Stephen J. Gould and Richard Lewontin, "The Spandrels of San Marco and the Panglossian Paradigm: A Critique of the Adaptationist Programme," *Proc. Roy. Soc. London B,* vol. 205, pp. 581–598.
23. François Jacob, *Le Jeu des possibles,* Fayard, Paris, 1981 (translated as *The Possible and the Actual,* University of Washington Press, Seattle, 1982).
24. F. Jacob, "Evolution and Tinkering," *Science,* vol. 196, 1977, pp. 1161–1166.
25. Such as the theory of Richard Goldschmidt, *The Material Basis of Evolution,* Yale University Press, New Haven, Conn., 1940 (reprinted with a preface by S. J. Gould, 1982).
26. Mary-Claire King and Allan C. Wilson, "Evolution at Two Levels in Humans and Chimpanzees," *Science,* vol. 188, 1975, pp. 107–116.
27. For a review of the early work in the direction of this hypothesis, see Allan C. Wilson, Steven S. Carlson, and Thomas J. White, "Biochemical Evolution," *Ann. Rev. Biochem.,* vol. 46, 1977, pp. 573–639.
28. For example, the ectopic expression of the Hox A7 gene in mice causes the ap-

pearance of a cervical vertebra—the "proatlas"—which disappeared at the time of dinosaurs: Michael Kessel, Rudi Balling, and Peter Gruss, "Variations of Cervical Vertebrae after Expression of a Hox 1.1 Transgene in Mice," *Cell,* vol. 61, 1990, pp. 301–308.

29. Stephen Jay Gould, *Ontogeny and Phylogeny,* the Belknap Press of Harvard University Press, Cambridge, Mass., 1977; Hervé Le Guyader, "Et quand la biologie moléculaire redécouvre la récapitulation . . ." in *Histoire du concept de recapitulation: ontogenèse et phylogenèse en biologie et sciences humaines,* Paul Mengal (ed.), Masson, Paris, 1993, pp. 111–129.
30. Jerry A. Coyne, "Genetics and Speciation," *Nature,* vol. 355, 1992, pp. 511–515.
31. A concern not expressed in Mayr's later writings. See "On the Evolutionary Synthesis and After," in *Toward a New Philosophy of Biology: Observations of an Evolutionist,* the Belknap Press of Harvard University Press, Cambridge, Mass., 1988, essay 28, pp. 525–554.

Acknowledgments

I would like to thank my colleagues—scientists and historians—for our many discussions, and in particular Charles Galperin and Jean-Paul Gaudillière. I would also like to thank Madame G. Morange for her invaluable help in preparing the manuscript.

Index

Index